River-Crossing Operations

Fredonia Books
Amsterdam, The Netherlands

River-Crossing Operations

by
U. S. Army
U. S. Marine Corps

ISBN: 1-4101-0783-3

Fredonia Books
Amsterdam, The Netherlands
http://www.fredoniabooks.com

Field Manual 90-13
Marine Corps Warfighting Publication 3-17.1

*FM 90-13
MCWP 3-17.1
Headquarters
Department of the Army
Commandant, US Marine Corps
Washington, DC, 26 January 1998

River-Crossing Operations

TABLE OF CONTENTS

PREFACE

Field Manual (FM) 90-13 describes how divisions and brigades conduct river crossings. It shows the relationship to corps operations, where appropriate, and includes details for lower echelons to support the brigades. It provides doctrine, tactics, techniques, and procedures (TTP) in one reference to accomplish this special operation.

The corps assigns missions and provides the necessary support and equipment. The divisions normally assign bridgehead objectives and control movement across the river. The brigades are the bridgehead forces that execute the crossings, either independently or as elements of a larger force.

River-crossing skills and knowledge are highly perishable. As with many other tactical operations, they require constant practice in planning and execution. There are relatively few opportunities to train with the frequency needed to keep a high degree of proficiency in this tough operation. For that reason, this manual includes considerable detail on techniques and procedures.

A river crossing is a special operation in that it requires specific procedures for success because the water obstacle prevents normal ground maneuver. It demands more detailed planning and technical support than normal tactical operations. It also features specific control measures to move the force across a water obstacle. This obstacle may be a river, a lake, or a canal. Unlike other obstacle types, the water obstacle remains effective during and after the crossing operation. See *FM 90-13-1* for other counterobstacle operations.

As in the past, the United States (US) Army conducts river crossings within the context of its basic doctrine. This manual applies the current Army-operations doctrine described in *FM 100-5* to river crossings. It incorporates recent developments in command and control (C^2) for command-post (CP) facilities and the military decision-making process. It also aligns US doctrine more closely with ongoing standardization efforts in the North Atlantic Treaty Organization (NATO).

Appendix A contains an English to metric measurement conversion chart.

The proponent of this publication is HQ TRADOC. Send comments and recommendations on *Department of the Army (DA) Form 2028* directly to Commander, US Army Engineer School, ATTN: ATSE-TD-D-WC, Fort Leonard Wood, MO 65473-6650.

This publication implements the following international agreement: *Standardization Agreement (STANAG) 2395, Edition 1, Opposed Water Crossing Procedures.*

Unless this publication states otherwise, masculine nouns and pronouns do not refer exclusively to men.

CHAPTER 1
Concepts

GENERAL

The purpose of any river crossing is to project combat power across a water obstacle to accomplish a mission. A river crossing is a unique operation. It requires specific procedures for success because the water obstacle prevents normal ground maneuver. It also requires detailed planning and control measures and different technical support than other tactical operations require. The nature and size of the obstacle, the enemy situation, and available crossing assets limit the tactical commander's options.

The challenge is to minimize the river's impact on the commander's ability to maneuver. The force is vulnerable while crossing, as it must break its movement formations, concentrate at crossing points, and reform on the far shore before continuing to maneuver. The tactical commander cannot effectively fight his force while it is split by a river. He must reduce this vulnerability by decreasing his force's exposure time. The best method is to cross rivers in stride as a continuation of the tactical operation, whether in the offense or retrograde. Only as a last resort should the force pause to build up combat power or crossing means before crossing. This chapter introduces river-crossing operations by discussing the characteristics of this special, difficult, and dangerous task.

TYPES OF CROSSINGS

Units expected to conduct a river crossing anticipate and plan for it in advance. All river crossings require detailed planning. The planning requirements and engineer technical support are similar, whether the crossing is hasty, deliberate, or retrograde.

HASTY

A hasty river crossing is a continuation of an attack across the river with no intentional pause at the water to prepare, so that there is no loss of momentum. This is possible when enemy resistance is weak and the river is not a severe obstacle.

A hasty river crossing is preferable to a deliberate crossing. A hasty river crossing features decentralized control at the brigade level. The brigade may use organic, existing, or expedient crossing means, but additional support from the division or corps is often necessary due to the bridge companies being controlled at corps level. That support is only available when those headquarters have taken purposeful action to position the assets at the right time and place to make a brigade hasty crossing feasible. Coordination for support must be made early in the planning process.

Small gaps that prohibit vehicles from self-bridging are encountered more frequently than large gaps that require extensive bridging. Each maneuver force should task-organize itself with organic mobile crossing assets that enable it to install bridges quickly, cross small gaps, and recover the bridges for future crossings. Follow-on bridges, such as the medium-girder bridge (MGB), may need to be positioned before assault bridges are removed at these minor gaps. The two types of hasty crossings are the dry- and wet-gap crossings.

Hasty Dry-Gap Crossing

Antitank (AT) ditches and craters are normally what maneuver forces encounter as a dry-gap-crossing obstacle. Dry riverbeds may also present a crossing problem. Maneuver forces can use the M9 armored combat earthmover (ACE) to push down the sides of ditches or to fill in craters. Substantial fill material placed in the dry gaps allows the passage of combat tracked vehicles. The crossing site can be improved and maintained for wheeled-traffic use by follow-on forces.

The armored vehicle-launched bridge (AVLB) is particularly suited for spanning streambeds, AT ditches, craters, canals, partially blown bridges, and similar obstacles. It can be launched and recovered in less than 5 minutes. The AVLB, like the M9 ACE, is organic to combat engineer companies for use in hasty crossings of short gaps. The AVLB should be left in place across the gap only as long as it takes to cross the maneuver unit it is traveling with, then replaced with other fixed bridging, if necessary.

Hasty Wet-Gap Crossing

The depth and width of the wet gap, bank conditions, and the current's velocity will determine if the maneuver force can cross by fording, swimming, or employing the AVLB or if other bridging assets are required. Identifying wet gaps early and deploying the required resources allow hasty crossings of known or anticipated gaps to occur. Two factors should be considered when swimming vehicles through wet gaps—the current's velocity and the bank conditions.

Because vehicles drain rapidly when exiting, initially firm banks tend to deteriorate rapidly from multiple uses of the same exit point. The existence of mud or surface irregularities further degrades the percent of the slope that the swimming vehicle can overcome. When selecting a fording site in a wet-gap crossing, the depth of the water is the most significant factor. The depth of the water in one crossing area may change due to bottom surface mud or irregularities (boulders or pot holes). The AVLB is ideally suited to allow hasty wet-gap crossings, requiring only that the supported maneuver force eliminate enemy direct and observed indirect fires. The crossing means will need to be replaced by other bridging assets as soon as possible to allow the AVLB to remain with its supported unit.

If possible, the force crosses the water obstacle at multiple points across a broad front. It makes the crossing as soon as its elements reach the obstacle, whether by day or night. As the bulk of the force crosses the water, minimum forces remain to secure the crossing sites.

Expedient crossing means may be used if readily available and can be transported to the crossing site. The reconnaissance party should note material or existing features that could be used as expedient crossing devices. These include culvert pipe, lumber or cut timber, or war-damaged equipment. The pipe fascines system (PFS), which consists of bundles of 8-inch, high-density, plastic pipes chained together, can fill gaps up to 9 meters deep and support up to 70 tons. The PFS is transported by an AVLB after the bridge is downloaded and emplaced into the gap.

A well-practiced standing operating procedure (SOP) reduces the necessary planning and preparation time. A concise order, clearly articulating the commander's intent, allows exploitation wherever subordinate units successfully force a crossing. When possible, advance elements seize existing crossing means intact and ahead of the main body.

When facing negligible or light enemy resistance on both banks, the force does not have to clear all enemy forces from the river to conduct a hasty crossing. It capitalizes on the speed of the crossing and the limited ability of the enemy to effectively oppose the crossing.

DELIBERATE

A deliberate river crossing is conducted when—

- A hasty crossing is not feasible.
- A hasty crossing has failed.

Opposition from a strong defending enemy can require a deliberate crossing. A deliberate river crossing is an attack across the river after a halt to make the detailed preparations necessary to ensure success. It is characterized by—

- A significant water obstacle.
- Strong enemy resistance.
- The necessity to clear entry and/or exit banks of enemy forces.

A deliberate river crossing involves the following:

- Centralized division planning and control.
- Thorough preparations, to include the time to perform extensive reconnaissance and full-scale rehearsals, develop alternate traffic routes, and stockpile logistics.
- The massing of forces and crossing equipment.

The deliberate river-crossing organization is as follows:

- An assault force that seizes the far-shore objective and eliminates direct fire on the crossing site.
- A maneuver-support force that consists of corps combat engineers, bridge companies, military police (MP), and chemical units which provide crossing means, traffic control, and obscuration.
- A bridgehead force that attacks from the far-shore objective to secure the bridgehead, eliminating direct fire and observed indirect fire on the crossing area.

Once the river crossing is complete (bridgehead line is secured), a breakout force crosses the river behind the bridgehead force and attacks out of the bridgehead. This force is normally not a part of the unit that conducted the river crossing.

The two types of deliberate crossings are wet- and dry-gap crossings.

Deliberate Wet-Gap Crossing

The deliberate wet-gap crossing is divided into the following three phases: assault, rafting, and bridging. These phases may occur in sequence or concurrently. The objective in deliberate wet-gap crossings is to project combat power to the exit bank at a faster rate than the enemy can concentrate forces for a counterattack. To do this, the commander may elect to first construct rafts for nonswimming vehicles while swimming the fighting vehicles across. Bridge construction is started when observed indirect fire has been eliminated. If the tactical situation allows the elimination of the rafting phase, bridging efforts should begin immediately. This may be a suitable option considering the high speed of employing systems like the ribbon bridge.

Deliberate Dry-Gap Crossing

Deliberate dry-gap crossings are generally determined by the strength of the enemy's defenses or the magnitude of the gap. If possible, using the M9 ACE or the AVLB is preferred. The MGB, the Bailey bridge, the M4T6 dry-span bridge and, in the near future, the heavy dry-support bridge (HDSB) are used to span larger dry gaps. These assets are labor-intensive and expose personnel to enemy fire during construction but provide stable gap-crossing support for continuous operations.

RETROGRADE

A retrograde river crossing is a movement to the rear across a water obstacle while in contact with the enemy. The forces conducting the crossing establish a defense on the exit bank or continue the retrograde to the defensive positions beyond the water obstacle. A retrograde river crossing features centralized planning and control because of the limited crossing means. It has the same amount of detailed planning as for a deliberate offensive crossing. Failure of a retrograde crossing may lead to losing a significant amount of friendly forces.

CROSSING FUNDAMENTALS

River-crossing fundamentals are the same for all river crossings, but their application varies. For example, traffic control is a key fundamental. The commander maintains it in a hasty crossing by using the unit's SOP and a fragmentary order (FRAGO). In a deliberate crossing, he uses a traffic-control organization, such as the MP, that implements a detailed movement plan. Crossing fundamentals must be applied to ensure success when conducting a river crossing. These fundamentals include—

- Surprise.
- Extensive preparation.
- Flexible plan.
- Traffic control.
- Organization.
- Speed.

SURPRISE

The range and lethality of modern weapons allow even a small force to defeat a larger exposed force caught in an unfavorable position. A river provides this possibility by—

- Limiting a force to a small number of crossing sites.
- Splitting the force's combat power on separate banks.

- Exposing the force to fires while on the water.

Surprise minimizes these disadvantages; forces that fail to achieve surprise may also fail in the crossing attempt.

A deception plan is a key element of surprise. It reinforces the enemy's predisposition to believe that the force will take a particular course of action (COA). The enemy usually expects a crossing; however, it does not know where or when. A deception plan that employs reconnaissance, site preparations, force buildup, and preparatory fires at a time or location other than the intended crossing area may delay an effective enemy response to the true crossing.

The usual operations security (OPSEC) measures are also important. Commanders enforce camouflage, noise, thermal, electromagnetic, and light discipline. In particular, commanders closely control movement and concealment of river-crossing equipment and other obvious river-crossing preparations. Despite modern intelligence-gathering technology, the skillful use of night, smoke, fog, and bad weather for obscuration is still effective.

EXTENSIVE PREPARATION

Comprehensive intelligence of the enemy's composition and disposition and crossing-area terrain must be developed early, since planning depends on an accurate and complete intelligence picture.

Supporting forces, which typically include engineer battalions, bridge companies, air-defense batteries, smoke-generation companies, and MP companies, must link up early. They immediately begin crossing preparations and are available to train the crossing force during rehearsals.

Commanders plan and initiate deceptive operations early to mask the actual preparation. These operations should conceal both the time and location of the crossing, so they begin before and continue throughout the preparation period.

Work necessary to improve routes to handle the traffic volume of the crossing operation should occur early enough not to interfere with other uses of the routes. This requires a detailed traffic plan carefully synchronized with the deception plan.

Full-scale rehearsals are essential to clarify roles and procedures, train personnel, inspect equipment, develop teamwork, and ensure the unity of effort.

FLEXIBLE PLAN

Even successful crossings seldom go according to plan. A flexible plan enables the crossing force to adapt rapidly to changes in the situation during execution. It allows the force to salvage the loss of a crossing site or to exploit a sudden opportunity. A flexible plan for a river crossing is the result of thorough staff planning, not chance. Such a plan features—

- Multiple approach routes from assembly areas (AAs) to crossing sites.

- Lateral routes to redirect units to alternate crossing sites.

- Alternate crossing sites and staging areas to activate if enemy action closes the primaries.

- Crossing equipment held in reserve to replace losses or open alternate sites.

TRAFFIC CONTROL

A river is a significant obstacle that slows and stops units, thus impeding their ability to maneuver. Units are restricted to moving in column formations along a few routes that come together at the crossing sites. Traffic control is essential to cross units at the locations and in the sequence desired. Maximum crossing efficiency is achieved through traffic control. It also prevents the formation of targets that are susceptible to destruction by artillery or air strikes. In addition, effective traffic control contributes to the flexibility of the plan by enabling commanders to change the sequence, timing, or site of crossing units. The traffic-control organization can switch units over different routes or hold them in waiting areas as directed by the tactical commander.

ORGANIZATION

Commanders use the same C^2 nodes for river crossings as they do for other operations. These nodes, however, take on additional functions in river crossings. For this reason, commanders specify which nodes and staff positions have specific river-crossing planning and control duties. This may require some temporary collocation of headquarters cells (or individual augmentation) and an increase in communications means.

The tactical commander organizes his units into assault, maneuver-support, and bridgehead forces. He organizes support forces consisting of engineer, MP, and chemical units, as well as other combat-support units, into a crossing organization. This organization reports to the tactical commander's controlling headquarters. Since this is a temporary grouping, procedures that the controlling headquarters establishes must be clear, simple, and rehearsed by all elements to ensure responsive support of the plan and the unity of command.

Terrain management is an integral part of the crossing operation. The controlling headquarters assigns space for support forces to work on and for assault forces to concentrate on before crossing. Otherwise, they interfere with each other and become lucrative targets for indirect fires and enemy air attacks.

SPEED

A river crossing is a race between the crossing force and the enemy to mass combat power on the far shore. The longer the force takes to cross, the less likely it will succeed, as the enemy will defeat, in detail, the elements split by the river. Speed is so important to crossing success that extraordinary measures are justified to maintain it. The commander must allow no interference with the flow of vehicles and units once the crossing has started.

CHAPTER 2
Terrain and Enemy

GENERAL

Commanders maneuver their forces into positions of advantage over the enemy. Engineers analyze the terrain to determine the maneuver potential, ways to reduce natural and enemy obstacles, and how they can deny freedom of maneuver to the enemy by enhancing the inherent obstacle value of the terrain.

ESTIMATE OF THE SITUATION

Commanders and staffs develop estimates of the situation, described in *FM 101-5*, during the military decision-making process. Terrain and enemy aspects that are applicable to estimates for river-crossing operations are discussed in this chapter. Much of this information directly applies to the intelligence preparation of the battlefield (IPB). Refer to *FM 34-130* for more information on the IPB process.

TACTICAL REQUIREMENTS

Although terrain characteristics have a strong influence, tactical requirements ultimately determine the location of the crossing site(s). River conditions must allow the employment of available crossing means and the tactics required for the operation.

The far-shore terrain must support mission accomplishment; otherwise, crossing the river there serves little purpose. Crossing sites must also support the rapid movement of units to the far shore, or the enemy can win the force buildup race. Commanders balance the tactical use of the far-shore terrain against technical crossing requirements at the river to determine suitable crossing locations.

Nearshore terrain must support initial assault sites, rafting and bridging sites, and the assembly and staging areas used by the force. Routes to and from the river must support the quantity of traffic that is necessary for the operation and for the sustainment of the force in subsequent operations.

The enemy's disposition of forces may limit options for the commander. Because the river physically splits his force, he should execute his crossing operation where the enemy is most vulnerable or least able to react. This gives the commander time to mass his force on the far shore before the enemy can concentrate against it.

TERRAIN

The engineer is the terrain expert. He must work closely with the Intelligence Officer (US Army) (S2) during the planning process to determine advantages and disadvantages the terrain gives to both friendly and enemy forces.

CHARACTERISTICS

Rivers form unique obstacles. They are generally linear and extensive and normally cannot be bypassed. Meandering bends in rivers provide far-shore defenders with opportunities for flanking fires and observation of multiple crossing sites. The combined-arms team, as normally configured for combat, needs special preparation and equipment to carry it across river obstacles. After the attacking force crosses the river, it remains an obstacle for all follow-on forces.

A formation cannot breach a river wherever desired, as it can with most field obstacles. Likely crossing sites can be few and equally obvious to both the attacker and defender.

A river provides excellent observation and fields of fire to both the attacker and defender. It exposes the force on the water and makes it vulnerable while entering and leaving the water. It is also an aerial avenue of approach, allowing enemy aircraft low-level access to crossing operations.

Force buildup on the far shore is a race between the defender and the attacker. The river can be an obstacle behind the initial assault force, allowing the enemy to pin and defeat it in detail while preventing rapid reinforcement.

MILITARY ASPECTS

Terrain analysis for a river crossing includes the following military aspects of terrain: observation, cover and concealment, obstacles, key terrain, and avenues of approach (OCOKA). However, many details are peculiar to river crossings. These details include the specific technical characteristics of the river as an obstacle.

CURRENT

The current of a river is a major limiting factor. It imposes limits on all floating equipment, whether rubber assault boats, swimming armored vehicles, rafts, or bridges. The current's velocity determines the amount of personnel/equipment each type of floating equipment can carry or if it can operate at all. Current affects the distance that the floating equipment will drift downstream. Therefore, commanders must either select an offset starting point upstream to reach a desired point on the far shore or take additional time to fight the current. High current velocities make

control of a heavy raft difficult; therefore, landings require skilled boat operators and raft commanders and more time.

Current causes water pressure against floating bridges. Bridge companies use boats or an anchorage system to resist this pressure. The higher the current the more extensive the anchorage system must be. Higher currents provide velocity to floating objects, which can damage or swamp floating equipment.

Current can be measured easily (for example, by timing a floating stick) but is normally not constant across the width of the river. Generally, it is faster in the center than along the shore. It is also faster on the outside of a curve than on the inside. A factor of 1.5 times the measured current should be used for planning purposes.

WATER MEASUREMENTS

The depth of the water influences all phases of a river crossing. If the water is shallow enough and the riverbed will support traffic, fording is possible. If the force uses assault boats and the water becomes shallow in the assault area, the force will have to wade and carry their equipment. Shallow water also causes difficulty for swimming vehicles, as the rapidly moving tracks can dig into a shallow bottom and ground the vehicle. The water must be deep enough to float bridge boats and loaded rafts on their crossing centerlines and deep enough in launch areas to launch boats and bridge bays. The depth of the water is not constant across a river. It is generally deeper in the center and in high-velocity areas. Either a bottom reconnaissance with divers or sounding from a reconnaissance boat is necessary to verify the depth.

The width of a river is a critical dimension for bridges (especially, when it determines how much equipment is necessary) and for rafts. The distance a raft must travel determines its round-trip crossing time, which in turn determines the force buildup rate on the far shore.

WATER CHANGES

A swell is the wave motion found in large bodies of water and near the mouths of rivers. It is caused by normal wave action in a larger body, from tidal action, or from wind forces across the water. A swell is a serious consideration for swimming armored vehicles and is less important for assault boats, heavy rafts, and bridges. Hydrographic data and local residents are sources of information on swells. Direct observation has limited use, as a swell changes over time with changing tide and weather conditions.

Tidal variation can cause significant problems. The depth and current of water change with the tide and may allow operations only during certain times. Tidal variation is not the same every day, as it depends on lunar and solar positions and on the current's velocity. Planners need tide tables to determine the actual variation, but they are not always available for rivers. Another tidal phenomenon found in some estuaries is the tidal bore, which is a dangerous wave that surges up the river as the tide enters. It seriously affects water operations. This reverse flow may require that float bridges be anchored on both sides.

Rivers may be subject to sudden floods due to heavy rain or thawing upstream. This will cause bank overflow, higher currents, deeper water, and significant floating debris. If the enemy possesses upstream flood-control structures or dams, it can cause these conditions also.

OBSTRUCTIONS

Most rivers contain sand or mud banks. They are characteristic of low-current areas along the shore and on the inside of the curves of a river, but they can be anywhere. Since they cause problems for swimming vehicles, assault boats, outboard motors, bridge boats, and rafts, troops must find them through underwater reconnaissance or sounding.

Rocks damage propellers, boats, and floating bridges and ground rafts. They cause swimming armored vehicles to swamp if the vehicle body or a track rides up on them high enough to cant the vehicle and allow water into a hatch or engine intake. They can also cause a fording vehicle to throw a track. Rocks are found by underwater reconnaissance or sounding.

Natural underwater obstructions and floating debris can range from sunken ships to wreckage and snags. The current in large waterways can carry significant floating debris, which can seriously damage boats and floating equipment. Usually, debris can be observed after flooding or rapidly rising waters. Underwater reconnaissance or bottom-charting sonar is the only way to locate underwater obstructions.

Man-made underwater obstacles can be steel or concrete tetrahedrons or dragon's teeth, wood piles, or mines. The enemy places them to deny a crossing area and designs them to block or destroy boats and rafts. Underwater reconnaissance or bottom-charting sonar can locate these obstacles.

Vegetation in the water can snag or choke propellers and ducted impellers on outboard motors and bridge boats. Normally, floating vegetation is not a significant problem. Thick vegetation beds that can cause equipment problems are found in shallow water and normally along the shore. As thick vegetation must extend to within 30 to 60 centimeters of the surface to hinder equipment, it can normally be seen from the surface.

THE FRIENDLY SHORE

Concealment is critical to the initial assault across the river. The assault force must have concealed access to the river. It must also have concealed attack positions close to the river from which to prepare assault boats. The overwatching unit prepares concealed positions along the friendly shore, taking full advantage of vegetation and surface contours. Overwatching units must be in position to engage the most likely enemy position(s) on the enemy shore.

Dominant terrain formed by hill masses or river bluffs provides direct-fire overwatch positions. If the dominant terrain is along the shore, it also covers attack positions, AAs, and staging areas. Air-defense (AD) sites should be located on terrain that dominates aerial avenues of approach (one of which is located along the river). When selecting a crossing site, consider the following:

- Dismounted avenues of approach that allow silent and concealed movement of assault battalions to the river.

- Concealed attack positions that are very close to the water along the dismounted avenue.

- Approaches from the attack positions to the water that have gradual slopes and limited vegetation to allow the assault force to carry inflated assault boats.

- Bank conditions that are favorable. Dismounted forces must be able to carry assault boats to the water, and engineer troops must be able to construct and operate rafts with little bank preparation.

- Road networks that feed the crossing sites and support the lateral movement of vehicles between sites. These road networks must be well constructed to carry large amounts of heavy vehicle traffic.

- Potential staging areas that can support large numbers of tracked and wheeled vehicles without continual maintenance.

- Helicopter landing zones (LZs) for embarkation of the assault force.

THE ENEMY SHORE

River meanders form salients and reentrant angles along the shore. A salient on the enemy shore is desirable for the crossing area, as it allows friendly fires from a wide stretch of the near shore to concentrate against a small area on the far shore and limits the length of enemy shore that must be cleared to eliminate direct fire and observation (see *Figure 2-1*).

Dominant terrain is undesirable on the enemy shore. Any terrain that permits direct or observed indirect fires onto crossing sites is key terrain. Friendly forces must control it before beginning the rafting or bridging phases.

Natural obstacles must be minimal between the river and the bridgehead objectives. River valleys often have parallel canals, railroad embankments, flood-control structures, swamps, and ridges that can impede more than the river itself. Obstacles perpendicular to the river can help isolate the bridgehead.

Exits from the river must be reasonably good without preparation. Initially, the bank should allow the assault force to land and dismount from the assault boats. This requires shallow banks with limited vegetation. The assault force also requires concealed dismounted avenues up from the river. Bank conditions must allow vehicles to debark from rafts and move up from the river. If banks require earthwork, at least one unimproved crossing site must allow the landing of earthmoving equipment. The most important far-shore requirement is a road network to carry high volumes of heavy vehicle traffic.

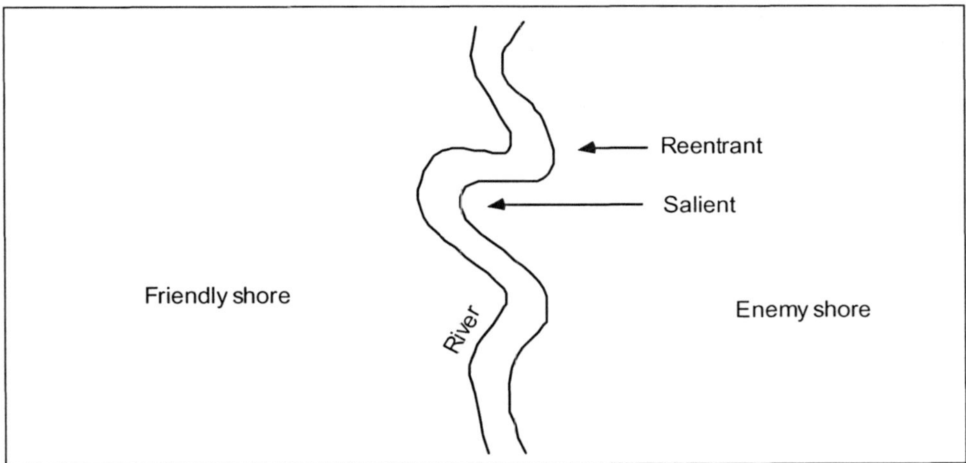

Figure 2-1. Salient and reentrant on the enemy shore

INTELLIGENCE

Detailed knowledge of the river and adjacent terrain is critical to both tactical planning and to engineer technical planning. The keys are early identification of intelligence requirements and an effective collection plan. Space-based imaging and weather systems can provide invaluable information to the terrain database. Multispectral imagery (MSI) from satellites can give the engineer terrain detachment a bird's-eye view of the area of operations. Satellite images, the largest 185 by 185 kilometers, can be used to identify key terrain and provide crossing locations. These images can provide information concerning the depth and turbidity of the river and can be used to identify the line of site for weapons and communications systems. With MSI products, prospective construction materials, the locations of existing crossing sites, and nearshore and far-shore road networks can be identified and exploited.

When the MSI is combined with satellite weather receivers, data processors, and the terrain database, it can be used to identify mobility corridors and establish floodplain trafficability. When these space systems are used together, the effects of the weather on terrain can be analyzed and used to develop decision-support products for the commander.

The terrain database is the starting point for obtaining terrain information. Hydrographic studies exist for most rivers in potential theaters of operation around the world. Many of these studies have sufficient detail for identifying feasible crossing sites. Modern information-collection and -storage technology permits frequent revision of existing data.

Engineer terrain detachments at corps and division maintain the terrain database and provide information in the form of topographic products. These products are used with other tools, such as computers and photography, to develop terrain intelligence for staff planners. The planners, in turn, determine initial crossing requirements and estimated crossing rates from their terrain analyses.

Early in the mission analysis, planners identify further terrain-intelligence needs for the crossing. They provide this to the Assistant

Chief of Staff, G2 (Intelligence) (G2) for inclusion in the intelligence-collection plan. The plan specifies that intelligence systems are used to gather essential terrain information for a more detailed analysis. Information on specific river segments and the surrounding terrain is obtained and verified by aerial and ground reconnaissance.

PRIORITY INTELLIGENCE REQUIREMENTS (PIR)

The following tactical and technical information is often PIR for executing a successful crossing:

- Enemy positions that can place direct or observed indirect fires on crossing sites and approaches.

- The location and type of enemy obstacles, particularly mines, in the water and on exit banks.

- The location of enemy reserves that can counterattack assault units.

- The location of enemy artillery that can range crossing sites, staging areas, and approaches.

- The location and condition of existing crossing sites.

- The width, depth, and velocity of the river.

- The condition and profile of the river's bottom.

- The height, slope, and stability of the bank.

- The condition of nearshore and far-shore road networks.

- Previous enemy tactics for defending water obstacles.

- Floodplain trafficability.

INFORMATION COLLECTION

Engineer units have the primary responsibility to collect the terrain information needed for river crossings. If the river is under friendly control, engineer units collect river, bank, and route information. If it is not, space- (satellite) or computer-based intelligence should be accessed, or maneuver units with attached engineer reconnaissance teams should conduct reconnaissance operations to obtain needed information. Engineer light diving teams obtain far-shore, nearshore, river-bottom, and underwater-obstacle information. Local inhabitants provide additional information about bridges, the flow of a river and the stability of its banks, road networks, ford sites, and other river conditions. Aviation assets can provide aerial and video reconnaissance to greatly enhance the IPB for river-crossing operations. Normal intelligence-collection assets develop the picture of the enemy's defense that is necessary for templating.

THREAT

Leaders who understand enemy tactics can defeat the enemy at the river for a successful crossing. Many potential enemies use doctrine from the former Soviet Union, making their tactics the most likely ones US forces must overcome during a crossing. Therefore, the discussion in the following paragraphs describes an opposing-force (OPFOR)-style defense and an attack at a river as the most likely threat. See *US*

Army Training and Doctrine Command (TRADOC) Pamphlet 350-14 for details on an OPFOR defense and *TRADOC Pamphlet 350-16* for OPFOR water crossings.

RIVER DEFENSE

The threat considers a water obstacle to be a natural barrier, enabling a strong defense on a wide front with small forces. Units must be prepared to conduct operations in a high level of mission-oriented protective posture (MOPP). The threat prefers to defend on a riverbank that is under its complete control. It can, however, defend forward or to the rear of a river. Its choice depends on the terrain, the forces available to it, and their strengths. The threat considers the defensive characteristics of the terrain. It weighs the severity of the obstacle, the effect of lost crossing sites, and the possibility of severed supply lines.

The threat may defend forward when the terrain is favorable, when it has sufficient reserve combat power, or when it plans to resume the offense immediately. When defending forward, it intends to defeat the crossing force before it reaches the river. The threat will place its defensive forces as far forward of the river as possible.

First-echelon regiments of a division in the main defensive belt forward of a river establish initial defensive positions 10 to 15 kilometers from the river. Second-echelon regiments occupy positions within a few kilometers of the river. These positions are astride major avenues of approach to block attacking forces so that a counterattack can destroy them.

When defending along a river, the threat places most of its forces as close to the exit bank as defensible terrain permits. Their mission is to protect the crossing sites and defeat the crossing force while it is divided by the river. The arrangement of defensive belts is similar to the defense forward of the river, except that the distance between first- and second-echelon regiments may be less. This concentrates more force to defeat assault forces on the exit bank.

Threat engineers destroy existing bridges and mine known crossing sites. They keep only a few sites open for the withdrawal of the predominantly amphibious security force. Threat engineers also emplace obstacles along approach and exit routes, including the riverbanks. As time and assets permit, they add obstacles such as floating mines and underwater obstructions to further disrupt crossing efforts.

First-echelon defensive forces maneuver to bring maximum defensive fire on the crossing force. These defensive forces engage the crossing force with all possible organic and support weapons at crossing sites while it is crossing. Their mission is to defeat the crossing force before it can establish a bridgehead.

Second-echelon battalions, astride major egress routes from the river, block assault elements so counterattacking forces can engage and destroy battalion or smaller assault elements. Second-echelon regiments occupy positions 4 to 5 kilometers behind the first echelon. They provide depth to the defense. It is from this area that the threat launches local counterattacks.

The threat undertakes a defense to the rear of a river when time or terrain precludes a defense forward of the river or on the exit bank. In this situation, security elements deploy on the exit bank to harass and disrupt the attacker's assault and support forces. These security elements delay the attacker to provide time to establish the main defense.

A significant threat capability against a river crossing is artillery. Therefore, if the S2 indicates that the threat has formed artillery groups (regimental artillery groups [RAGs], division artillery groups [DAGs], or Army artillery groups [AAGs]), then it has the capability to saturate crossing sites. In this case, it is not sufficient to eliminate the threat's observation of the river before building bridges, as the concentration of artillery fires can deny an entire bridging site without the necessity for observed fires.

CHAPTER 3
Command and Control

GENERAL

Unity of effort is established by the C^2 emplaced on the maneuver units, the crossing-force headquarters, and the supporting units. Unit organization and traffic control are fundamental to successful river-crossing operations. They enable the commander to apply the tactics discussed in *Chapters 5 and 6*. This chapter covers the techniques and procedures used to establish the crossing organization, maintain control of forces, and hand off responsibilities between echelons as the operation progresses.

ORGANIZATION

Division and brigade commanders organize their forces into assault, maneuver-support, bridgehead, and breakout forces for river-crossing operations. Assault forces seize the far-shore objective to eliminate direct fire on the crossing sites. Maneuver-support forces consist of corps combat engineers, bridge companies, MP, and chemical units that provide crossing means, traffic control, and obscuration. Bridgehead forces secure the bridgehead. Breakout forces cross the river behind the bridgehead forces and attack out of the bridgehead.

CONTROL ELEMENTS

Division and brigade commanders are responsible for crossing their formations. They organize their staffs and subordinate commanders to help them control the crossing (see *Table 3-1, page 3-2*). Division and brigade headquarters operate from echeloned CPs. They are the tactical, main, and rear CPs and provide the staff and communications support for planning and executing river crossings. The CPs may need some temporary augmentation or realignment of internal staff elements for the crossing. *Figures 3-1 and 3-2, pages 3-3 and 3-4*, show the necessary control elements for deliberate and retrograde river-crossing operations. Each of the control elements is discussed below.

DIVISION HEADQUARTERS

The division tactical CP (DTAC) controls the lead brigades' (bridgehead force) attack across the river, since this is the division's close fight. It may reallocate crossing means or movement routes to the river between brigades as the battle develops. The DTAC is the crossing-force headquarters.

The division main CP (DMAIN) prepares the river-crossing plan. It also directs the division's deep operations to isolate the bridgehead from enemy reinforcements and counterattacking formations. As a guide, the DMAIN displaces across the river after the division reserve. For division crossings, a traffic-control cell schedules, routes, and monitors traffic behind the lead brigades. The cell collocates with the DMAIN. The Assistant Chief of Staff, G4 (Logistics) (G4) provides the cell nucleus.

The division rear CP (DREAR) sustains the crossing for other division operations. Once the DMAIN displaces across the river, the crossing becomes a rear operation that the DREAR controls.

Table 3-1. CP tasks (deliberate crossing)

Phases / CPs	Advance to the River	Assault Across the River	Advance From the Exit Bank	Secure the Bridgehead Line	Continue the Attack
DTAC (crossing force's headquarters)	Coordinates the lead brigade's seizure of nearshore objectives	Coordinates the lead brigade's dismounted assault of the river to seize the far-shore objectives	Coordinates the lead brigade's seizure of exit-bank and intermediate objectives	Coordinates the lead brigade's seizure and securing of bridgehead objectives and prepares to cross the reserve brigade (breakout forces)	Controls the breakout force's attack out of the bridgehead and passes the crossing force's responsibilities to the DREAR
DMAIN	Coordinates deep operations to isolate the division's advance to the river	Coordinates deep operations to isolate the crossing area and the far-shore objectives	Coordinates deep operations to isolate exit-bank and intermediate objectives	Coordinates deep operations to isolate the bridgehead	Coordinates deep operations to isolate the enemy's attack against corps objectives
DREAR	Sustains the fight	Sustains the fight	Sustains the fight	Sustains the fight	Assumes the role of the crossing force's headquarters
BTAC	Coordinates the lead TF's seizure and securing of nearshore objectives	Coordinates the dismounted assault crossing of the river to secure the far-shore objectives	Coordinates the TF's attack to seize and secure exit-bank and intermediate objectives	Coordinates the TF's seizure and securing of bridgehead objectives	Prepares to reorganize and follow the breakout force's attack out of the bridgehead toward the division's deep objectives
BMAIN (crossing-area headquarters)	Moves into the crossing area to provide traffic control, crossing means, and obscuration	Coordinates assault crossing means for TFs dismounted and controls obscuration of the crossing sites	Controls follow-on TFs passing through the crossing area into attack positions	Controls the passage of the brigade's units through the crossing area and prepares to cross breakout forces	Passes crossing-area control to the supporting corps's engineer battalion

Crossing-Force Commander (CFC)

The division commander normally designates an assistant division commander (ADC) as the CFC to take charge of controlling the division crossing.

Crossing-Force Engineer (CFE)

A crossing division receives support from a CFE, who is normally the commander of an engineer group from the corps engineer brigade. He provides additional staff planners

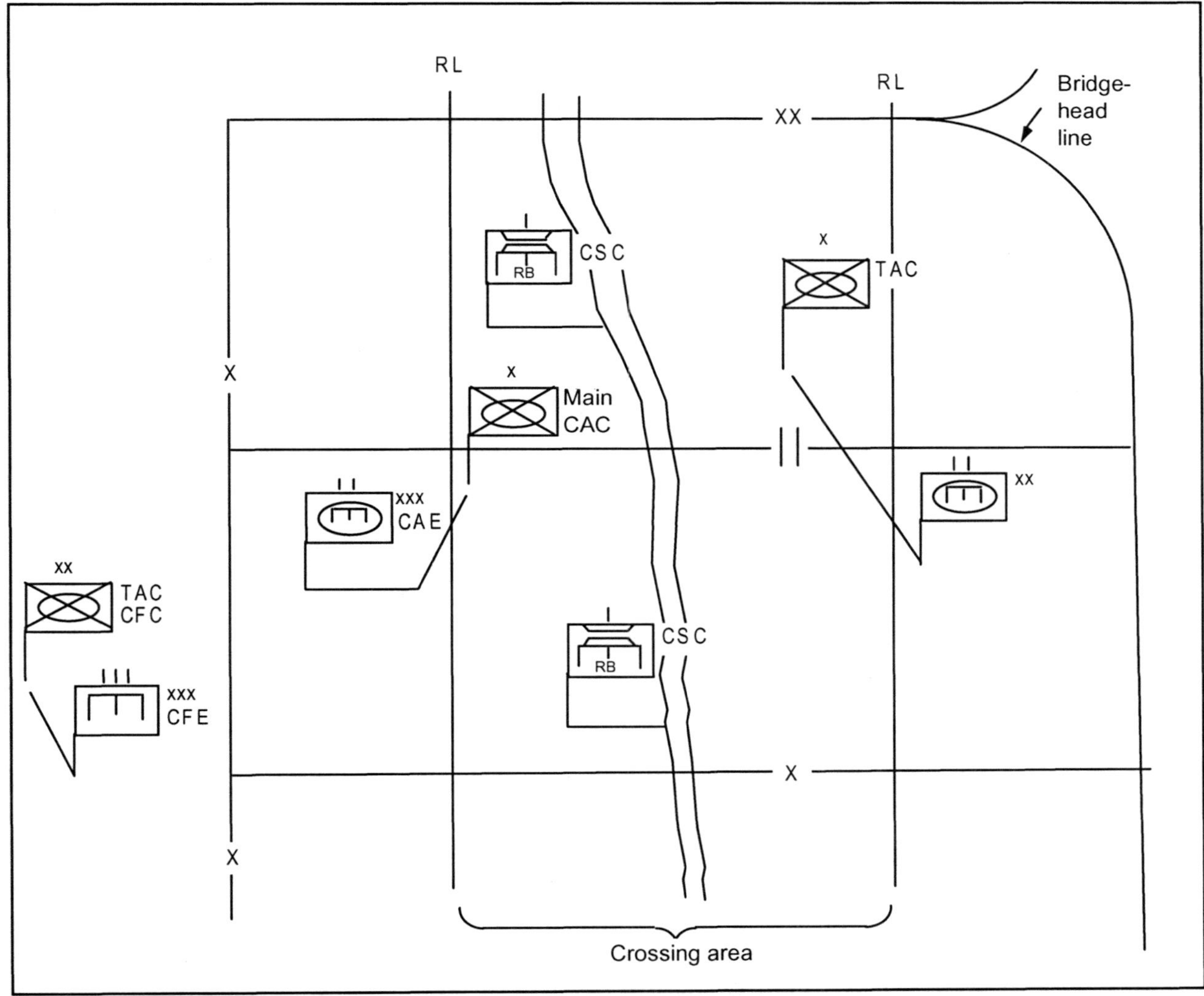

Figure 3-1. Control elements for a deliberate crossing (brigade focus)

for the CFC and coordinates engineer support to the crossing-area commanders (CACs).

BRIGADE HEADQUARTERS

Each brigade headquarters operates from echeloned CPs, the brigade tactical CP (BTAC), and the brigade main CP (BMAIN). The BTAC controls the advance to and the attack across the river. It displaces across the river as soon as practical after the assault across the river to control the fight for exit-bank, intermediate, and bridgehead objectives.

The BMAIN controls the crossing of the rest of the brigade. It prepares the brigade crossing plan and provides the staff nucleus to coordinate it. For brigade crossings, the Supply Officer (US Army) (S4), assisted by the supporting MP unit leader or engineers if available, organizes a small, temporary traffic-control cell collocated with the BMAIN.

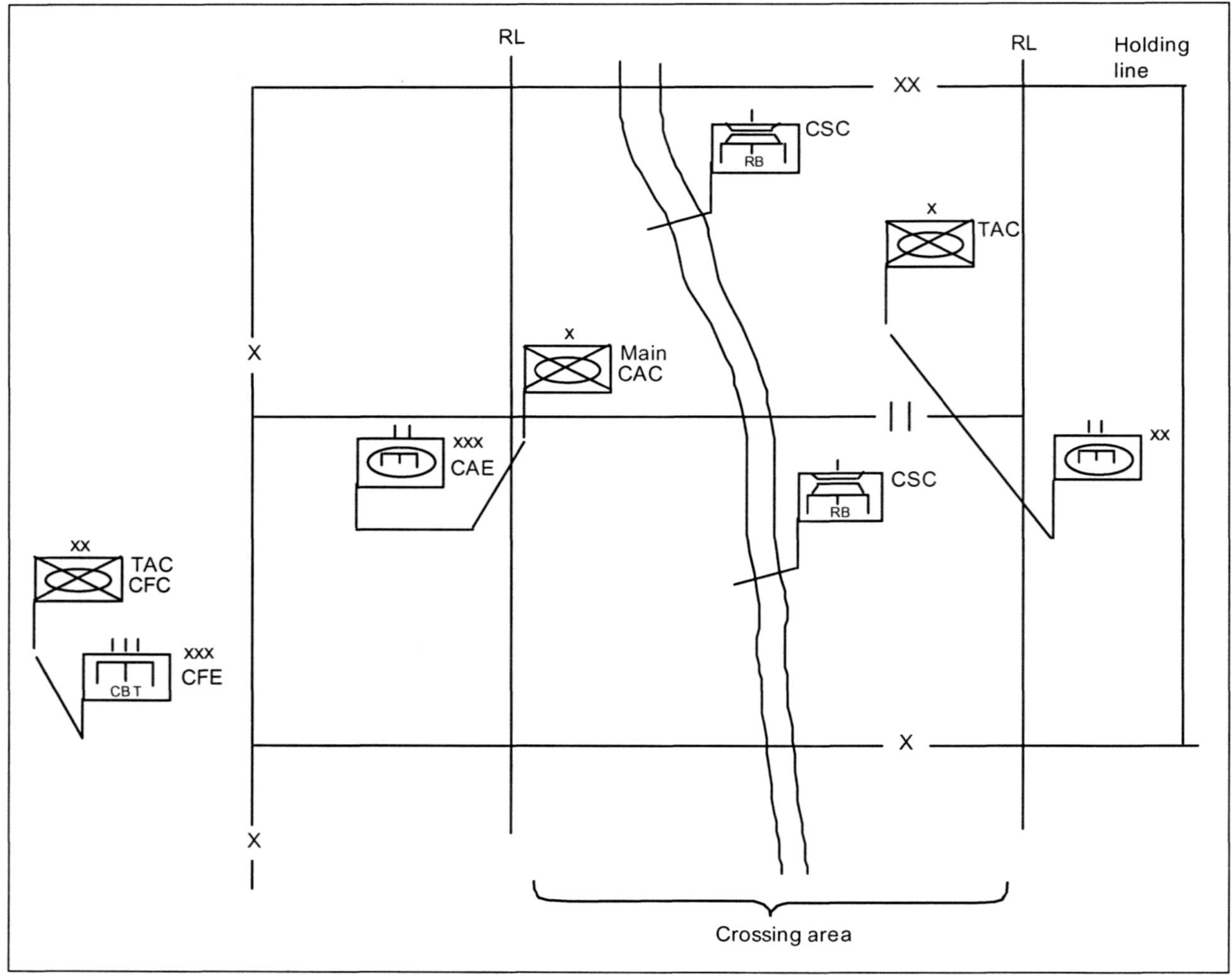

Figure 3-2. Control elements for a retrograde crossing (brigade focus)

Crossing-Area Commander

Once the lead battalions assault across the river and secure the far-shore objective, the crossing area is activated. The CAC, normally the brigade's executive officer (XO), controls the movement of forces inside the crossing area. The BMAIN controls the maneuver-support force that consists of corps engineers, bridge companies, and MP and chemical units. This leaves the brigade commander free to direct key activities while an officer who is directly responsible to him runs the crossing. The CAC controls—

• The movement and positioning of all elements transiting or occupying positions within the crossing area.

• Security elements at crossing sites.

• Maneuver-support forces, such as engineer, MP, and chemical units within the crossing area.

Crossing-Area Engineer (CAE)

Each forward brigade will normally be supported by a direct-support engineer battalion from the corps. The engineer battalion commander is responsible to the CAC for the engineer crossing means and sites. He informs the CAC of changes, due to technical difficulties or enemy action, that render a crossing means inoperable or reduce its capacity. He commands those engineers tasked to move the force across the river; they remain there as the attack proceeds beyond the exit-bank objectives. The division engineer battalion focuses on supporting the lead brigades at exit-bank, intermediate, and bridgehead objectives and is not normally involved in the river crossing.

Crossing-Site Commander (CSC)

Each crossing site has an engineer, either a company commander or a platoon leader, who is responsible for crossing the units sent to the site. Normally, the CSC is the company commander for the bridge unit operating the site. He commands the engineers operating the crossing means and the engineer regulating points (ERPs) at the call-forward areas for that site. He maintains the site and decides on the immediate action needed to remove broken-down or damaged vehicles that interfere with activities at the site. He is responsible to the CAE and keeps him informed on the status of the site.

Unit-Movement-Control Officer

Each battalion and separate unit commander designates a movement-control officer, who coordinates the unit's movement according to the movement-control plan. He provides staff planners with detailed information on the unit's vehicle types and numbers.

COMMUNICATIONS

Figures 3-3 and 3-4, pages 3-6 and 3-7, depict the communications networks within a crossing area. In the hasty-crossing example, a brigade making a supporting attack conducts a crossing with its normal slice of combat-support forces plus a corps bridge company. More assets are available from the division and corps in the deliberate-crossing example. Wire is the preferred means of communications in a river crossing when there is sufficient time to prepare it. The corps engineer battalion will establish wire communications with the nearshore crossing area according to the crossing plan.

CONTROL MEASURES

The commander uses control measures to delineate areas of responsibility for subordinates and to ease traffic control. *Figure 3-5, page 3-8,* illustrates the control measures described below.

RELEASE LINES (RLS)

As used in river-crossing operations, RLs are used to delineate the crossing area. RLs are located on both the far and near shores and indicate a change in the headquarters that is controlling movement. RLs are normally located within 3 to 4 kilometers of the river and on easily identifiable terrain features, if possible.

CROSSING AREAS

Crossing areas are controlled access areas that decrease congestion at the river. This permits swift movement of forces. Each lead brigade has a crossing area on both sides of the river that is defined by brigade boundaries and RLs. Crossing areas normally extend 3 to 4 kilometers on each side of the river, depending on the terrain and the anticipated battle.

WAITING AREAS

Waiting areas are located adjacent to the routes or axes of advance. Commanders use the following waiting areas to conceal

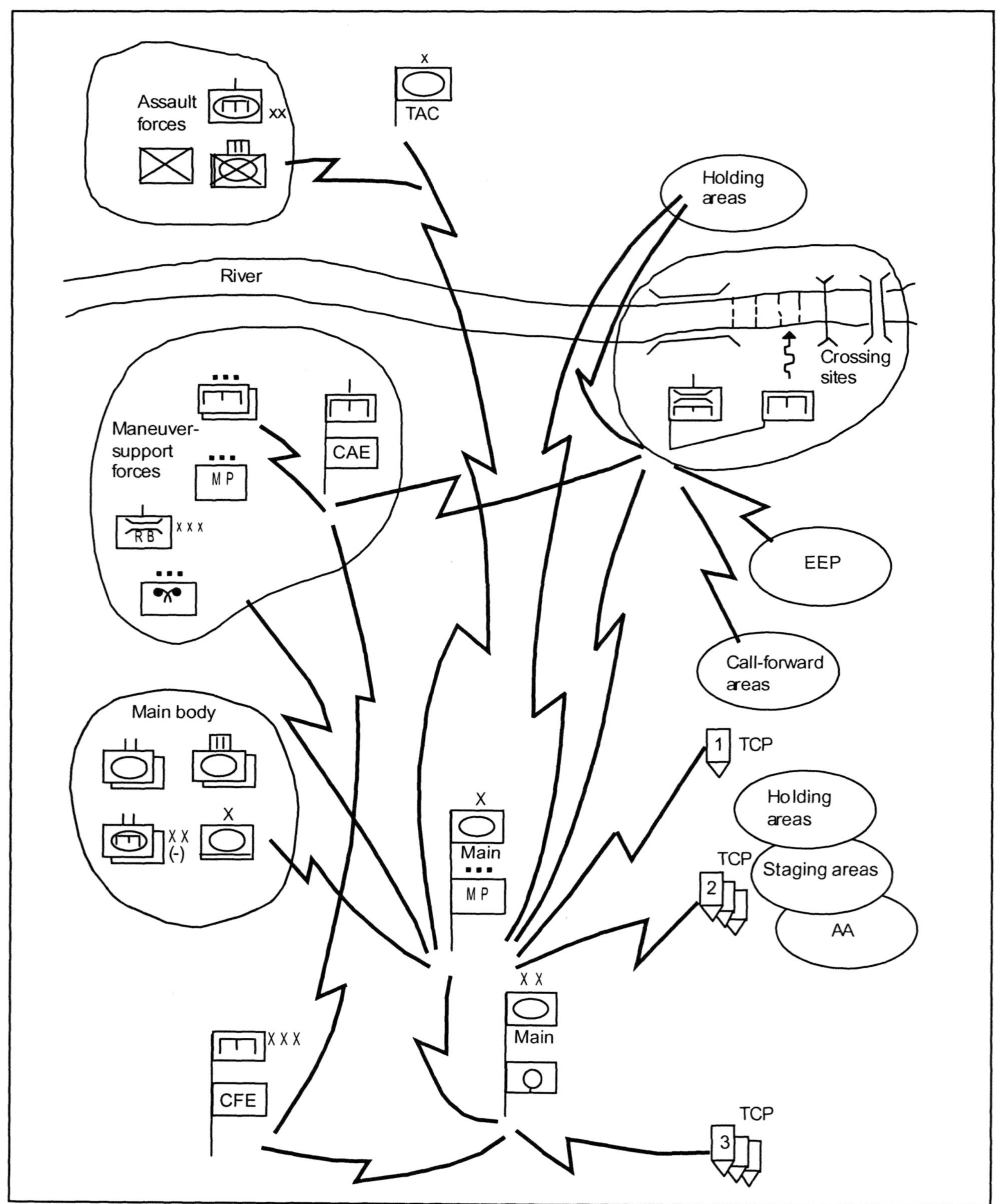

Figure 3-3. Communications for a deliberate crossing

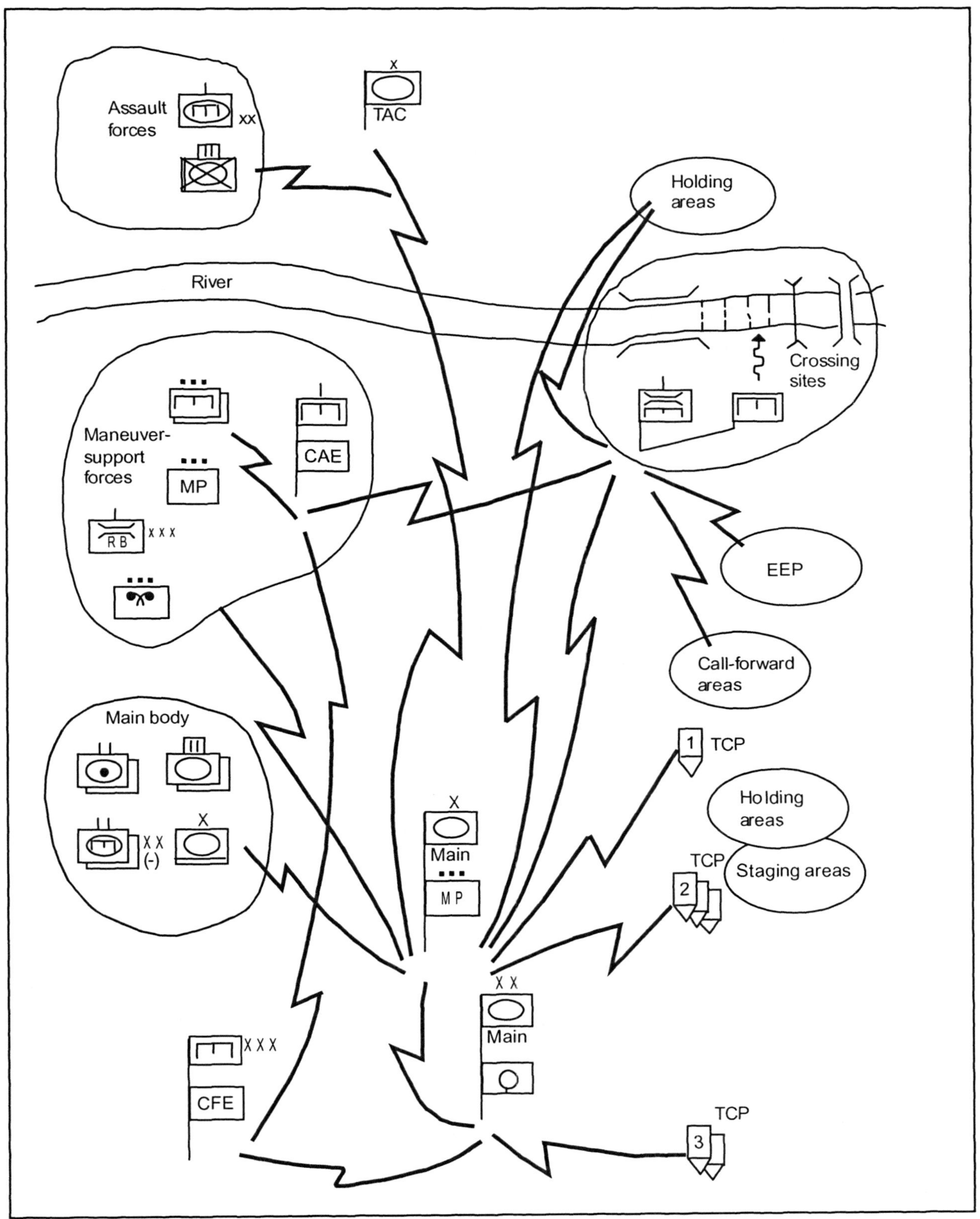

Figure 3-4. Communications for a hasty crossing

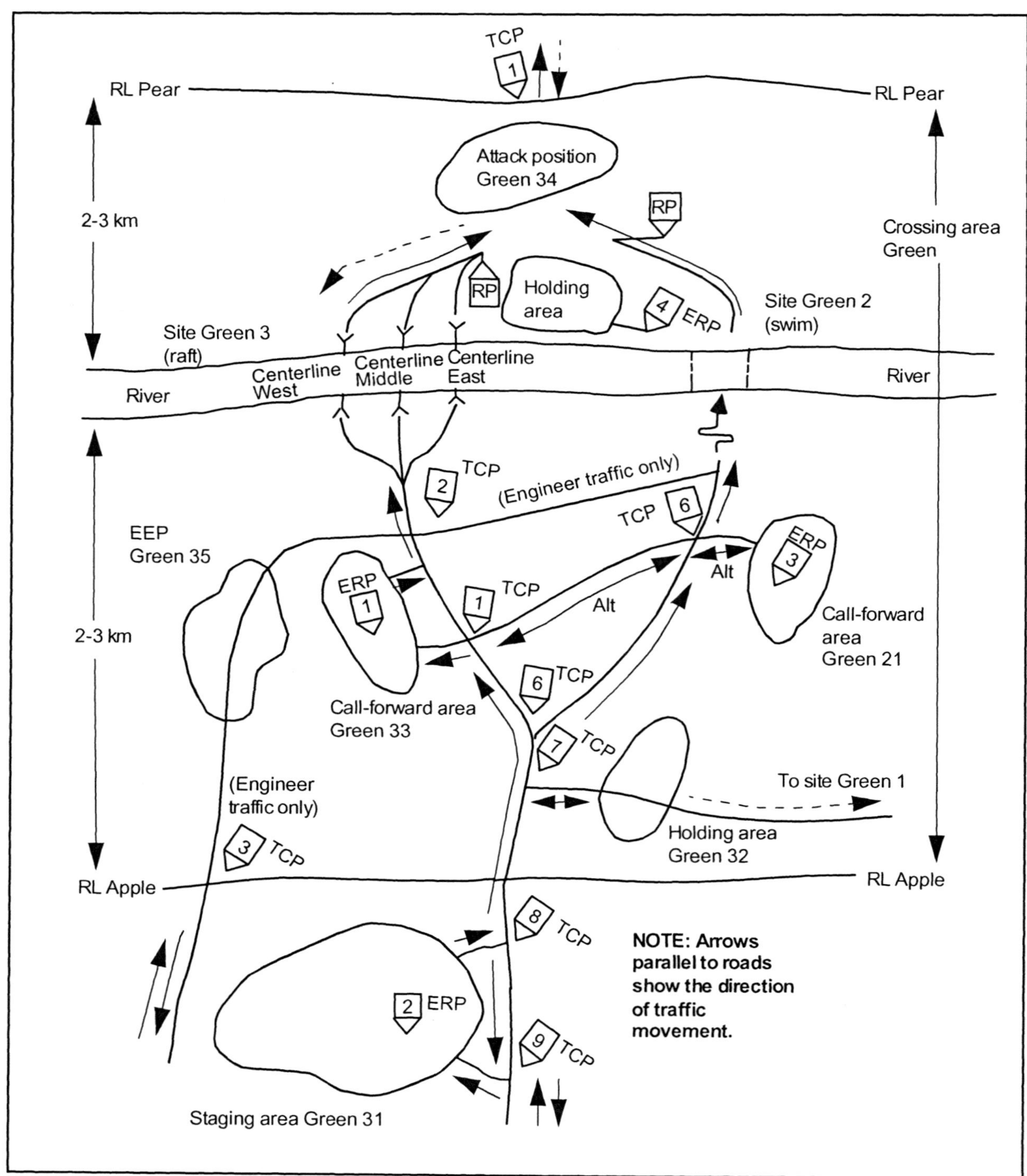

Figure 3-5. Control measures

vehicles, troops, and equipment while waiting to resume movement or to make final crossing preparations:

- Staging areas.
- Call-forward areas.
- Holding areas.
- Attack positions.
- AAs.

Staging areas

Staging areas are battalion-size waiting areas outside the crossing area where forces wait to enter the crossing area. The brigade traffic-control cell handles units' movement into staging areas. The CAC controls movement from the staging areas into the crossing areas. MP operate traffic-control posts (TCPs) at staging areas according to the crossing and traffic-circulation plans. They emplace temporary signs along the route from the staging area through the crossing area to guide convoys. Units make crossing preparations and receive briefings on vehicle speed and spacing in the staging areas. Staging areas—

- Are located to support the crossing concept.
- Are far enough back to permit the rerouting of the battalion along other roads or to alternate crossing sites.
- Are easily accessible from major routes.
- Have sufficient area for dispersing a battalion-size unit.
- Provide concealment.

Call-Forward Areas

Call-forward areas are company-size waiting areas located within the crossing area. Engineers use them to organize units into raft loads, or crews use them to make final vehicle swimming preparations. The CAC controls movement from the staging area to the call-forward area. The CSC directs movement from the call-forward area to the crossing site and on to the far-shore attack position. As a minimum, each CSC operates his own call-forward area. Call-forward areas—

- Are located to support the crossing plan.
- Are company size within the crossing area.
- Are easily accessible from routes.
- Are planned with a minimum of one per crossing site.
- Have ERPs collocated with them.
- Are used to organize units into raft loads.
- Are the final preparation areas before going to the crossing site.
- Are normally operated by engineers.

Holding Areas

Holding areas are waiting areas that forces use during traffic interruptions. Units move into these areas when directed by TCP personnel and disperse rather than stay on the roads. Holding areas are battalion size outside of the crossing area and company size within it. Far-shore holding areas are used to organize return traffic. MP and engineers, if available, operate holding areas according to the crossing and traffic-circulation plans. Established as needed on both sides of the river, holding areas—

- Are used as call-forward areas for return traffic from the far shore.
- Are located to support the crossing plan.
- Are easily accessible from routes.
- Have sufficient area for dispersion.
- Provide cover and concealment.
- Are defensible.
- Maximize traffic flow with minimum control.

Attack Positions

Attack positions are the last positions occupied or passed through by the assault echelon or attacking force before crossing the line of departure. Within the bridgehead, the attack position is the last position before leaving the crossing area or bridgehead line.

Assembly Areas

AAs are areas in which a force prepares or regroups for further action.

ENGINEER EQUIPMENT PARKS (EEPS)

EEPs are areas located a convenient distance from bridging and rafting sites for assembling, preparing, and storing bridge equipment and material. They are at least 1 kilometer from the river and hold spare equipment and empty bridge trucks that are not required at the crossing sites. EEPs should be located where they do not interfere with the traffic to the crossing sites and where equipment can be concealed and dispersed. Ideally, routes leading from EEPs to the crossing sites are not the same routes used by units crossing the river.

TRAFFIC-CONTROL POSTS

In river crossings, TCP personnel assist the crossing-area headquarters in traffic control by reporting and regulating the movement of units and convoys. TCP personnel relay messages between the crossing-area headquarters and moving units. The provost marshal identifies locations that need or require TCPs. MP or engineers, if available, operate TCPs on both banks of the river to control traffic moving toward or away from it. TCPs are additionally operated at major or critical crossroads and road junctions, staging areas, holding areas, and ERPs.

ENGINEER REGULATING POINTS

ERPs are technical checkpoints which are used to ensure that vehicles do not exceed the capacity of the crossing means. They help maintain traffic flow. Vehicles which will not be allowed to cross are removed so that they do not cause a traffic backup at the actual crossing site. Engineers man the ERPs and report to the CSC. TCPs are collocated with the ERPs to ensure that all vehicles clear the call-forward areas. An additional duty of ERP personnel is to give the drivers final instructions on site-specific procedures and other information such as speed and vehicle intervals. As a minimum, each crossing site requires an ERP at its own call-forward area. If sufficient engineer assets are available, ERPs may be established at far-shore holding areas to regulate rearward traffic.

CROSSING PLAN

The crossing plan is integrated throughout the division's and brigade's operation orders (OPORDs) and is as detailed as time permits. The crossing annex to the OPORD contains much but not all of the plan. It has the crossing overlay and the crossing synchronization matrix.

The crossing overlay shows the crossing areas, the crossing sites, the routes leading up to them from waiting areas, and all the control measures necessary for the crossing (see *Figure 3-6*). The crossing synchronization matrix is a tool to adjust the crossing plan as the battle develops. It shows crossing units in relation to their planned crossing times and locations. See *Appendix B* for an example matrix.

The *task organization paragraph and paragraph 5* of the OPORD contain the organization and command portions of the crossing plan. For more information on the development of the crossing plan, see *Chapter 4.*

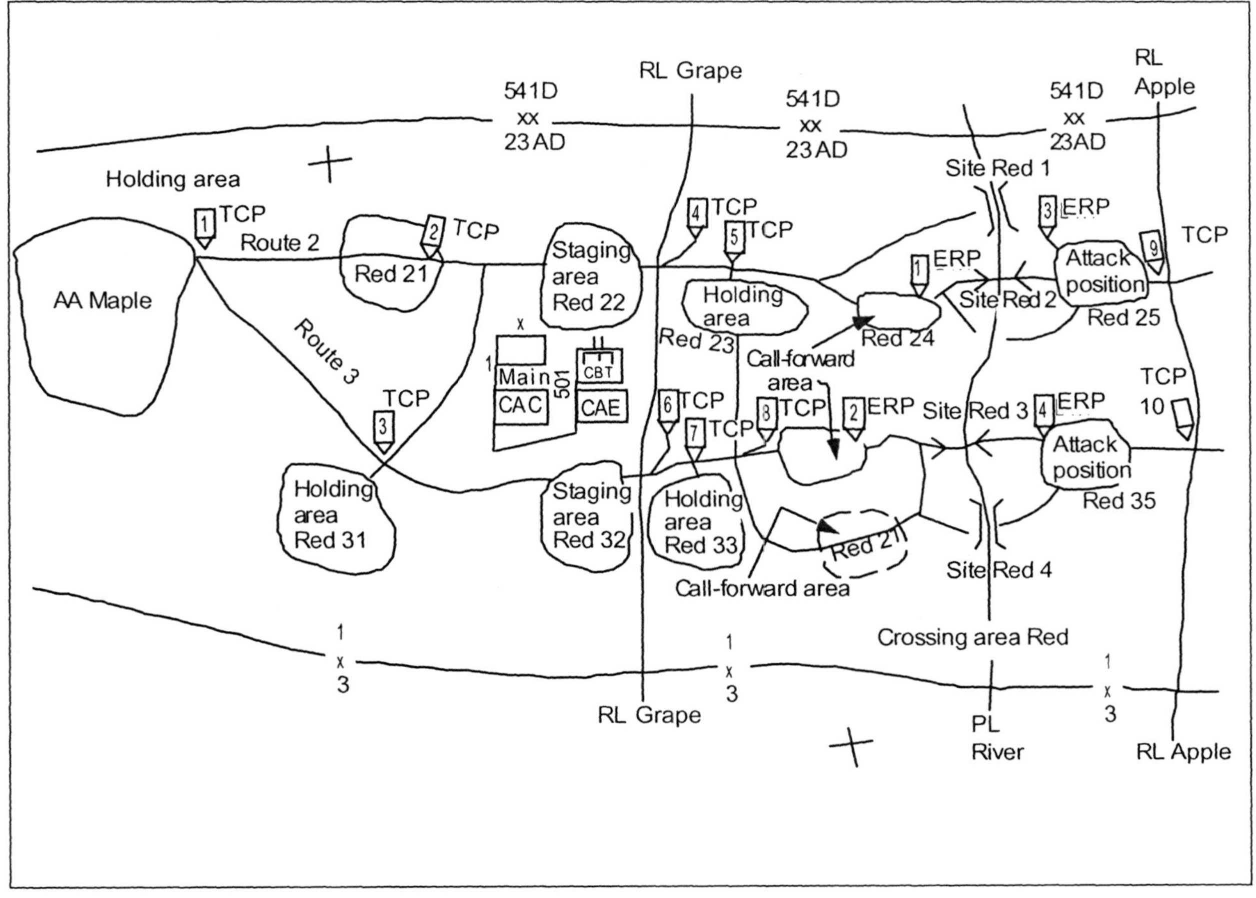

Figure 3-6. Crossing overlay

CROSSING CONTROL

Commanders use control measures to operate, delegate authority, and lead from any critical point during the river-crossing operation while synchronizing other critical actions throughout their area of operations.

ASSAULT ACROSS THE RIVER

Battalion task forces (TFs) conducting the assault across the river move to it under the direct control of their brigade commanders. The assault TFs using rubber boats 15 (RB15s) follow the procedures in *Chapter 8.* The brigade commander keeps the remainder of the brigade back from the river to avoid congestion. Elements not engaged in security or supporting the crossing occupy AAs and prepare for movement across the river.

CROSSING-AREA OPERATIONS

After the assault across the river, the brigade has an initial position on the far shore and is no longer fighting to seize the exit bank. The brigade needs its follow-on forces across as quickly as possible. The battalions can now cross without engaging in combat at the river. The brigade commander activates the crossing area to move forces rapidly and efficiently. The urgent need to get tanks across the river means the rafting stage often begins before terrain on the far shore is secure to the planned RL. Therefore, the crossing area is initially limited to the near shore. The first fighting vehicles swimming or rafting across under this circumstance have limited space to regroup before commitment to the fight.

As the initial battalions cross, they gain terrain to the necessary depth, and as control elements cross to the far shore, the brigade commander extends the crossing area out to the planned RL. Thereafter, units move completely through the crossing area under the CAC's control and exit it in a tactical move.

When rafting, the crossing flow for the follow-on units is generally from a staging area, through the call-forward area and crossing site into an attack position, and then on to a subsequent objective. While bridging, the flow is from a staging area, through the crossing site, and then out of the crossing area.

Figure 3-7 illustrates the traffic flow for a follow-on battalion TF during the rafting. This procedure avoids congestion close to the crossing site and helps maintain unit integrity while the battalion rafts. The battalion occupies staging area Green 31 and organizes an internal unit crossing order based on its mission on the far shore. When concurrently swimming and rafting vehicles of the same battalion, the swimming vehicles form up separate from nonswimming vehicles for movement to the crossing sites and reform into a tactical formation at the far-shore attack position. ERP personnel at the call-forward area check to determine the correct load classification and proper loading sequence for each vehicle. When instructed by the CAC, the battalion sends one company at a time (or the equivalent) from the staging area. TCP personnel guide the company's movement en route to a call-forward area where it comes under the movement control of the CSC.

In the call-forward area at site Green 33, ERP personnel organize individual vehicles into raft loads. They guide the raft loads down to the raft centerlines as the CSC directs. In the call-forward area at site Green 21, vehicle crews make the final vehicle swimming preparations. ERP personnel send the vehicles down to the swimming site when directed by the CSC.

Vehicles remain under the control of the CSC until they are on the far shore. There they proceed to attack position 6, where they regroup as a company/team. When ready, the TF commander, under the tactical control of the brigade commander, controls the movement of the vehicles.

During bridging operations, the CAC normally directs the follow-on battalions to move in company serials from the staging area. Each serial moves down to the bridge site, crosses the river, and continues on to the attack position. The CAC directs an interval between serials that keeps continuous traffic across the bridge without gaps or traffic jams. A call-forward area remains established in the event that the bridge becomes damaged and units must be held until raft operations resume.

Units in the support-by-fire position on the near shore are already inside the crossing area when the crossing operation starts. They remain in this position until the CAC directs them to cross the river, and then they move directly to previously selected call-forward areas or start points (SPs) by company or platoon.

TRANSFER OF SUPPORT FORCES TO DIVISION

Once the bridgehead forces are across the river, the crossing sites are relatively secure. Since ground maneuver is no longer close to the crossing area, the operation at the river becomes predominantly a bridging

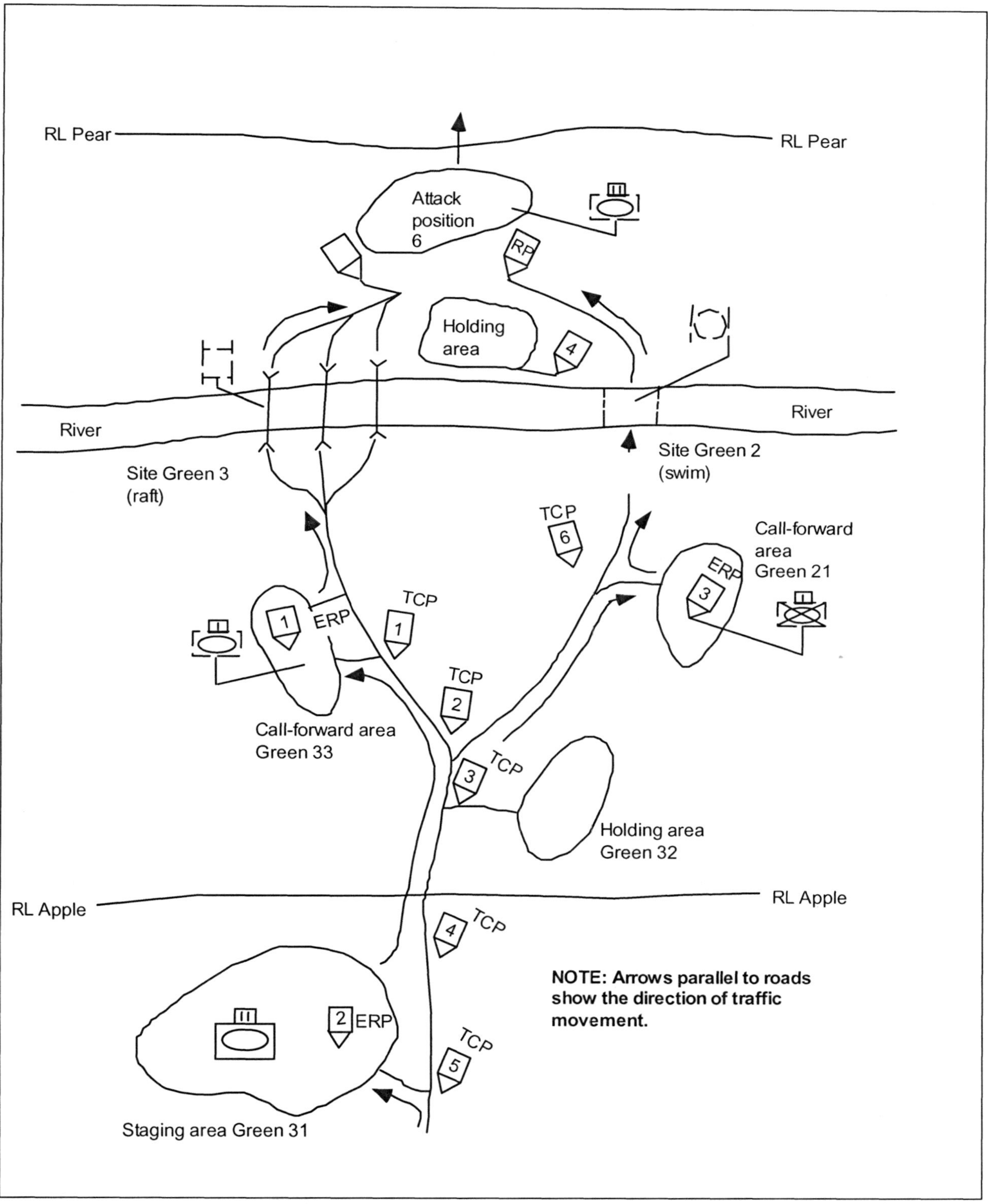

Figure 3-7. Follow-on-TF crossing during a rafting phase

and traffic-scheduling problem. The division headquarters moves the RL at the rear of the bridgehead force to the far shore. The crossing areas come under direct division control. As the ADC directs, the brigade commander turns over his crossing area to another officer, normally the CAE, who becomes responsible for the crossing area. The CAE then reports through the CFE to the ADC at the DREAR. The CAE's unit headquarters becomes the crossing-area headquarters.

MOVEMENT CONTROL

Movement control is vital to efficiently move units and material up to the crossing area in the sequence needed by the commander. The traffic-control cells at the division and brigade headquarters exercise movement control through TCPs. The division controls movement from its rear boundary up to the brigade rear, and the brigade controls movement from the rear boundary up to the bridgehead line.

The division transportation officer (DTO) develops the division movement plan according to the movement priorities that the Assistant Chief of Staff, G3 (Operations and Plans) (G3) and the G4 establish. The S4 prepares the brigade movement plan according to the priorities that the Operations and Training Officer (US Army) (S3) establishes. Each unit-movement officer, normally the battalion S4, provides the unit's vehicle information to the planning headquarters.

The movement plan normally consists of a traffic-circulation overlay and a road-movement table found in the movement annex to the division's or brigade's order.

RETROGRADE CROSSINGS

A retrograde river crossing has most of the same control features as an offensive crossing. The commander responsible for a crossing area has the same authority as he does in an offensive crossing. When a brigade establishes a defense along the river concurrent with the crossing, the commander coordinates crossing activities to avoid conflicts with defensive preparations. For this reason, the responsible officer and his staff should be familiar with both the delaying and defending commanders' tactical plans. He coordinates optimum use of crossing sites by delaying forces. As the delaying forces disengage, they must rapidly pass through the defending force and cross the river. The commander responsible for the crossing area reports to the division CP controlling the operation. If the main CP is forward of the river, C^2 is usually at the DREAR until the main CP displaces behind the river.

When the river is in the division's rear area at the start of the retrograde, the crossing begins as a rear operation. The senior corps engineer commander supporting the division becomes the CFE and establishes division crossing areas with corps engineer and MP units. He identifies engineer commanders, as directed by the commanding general, to quickly organize the crossing areas and initiate crossing control. These crossing areas correspond to the brigade boundaries planned by the G3 for the defense along the river.

Each brigade commander establishing a defense at the river appoints an XO to control the crossing area in his sector. When the river is in the brigade's sector at the start of the retrograde, this officer can immediately take charge and organize the crossing area. If the division initially organizes the crossing area through the CFE, it directs the defending brigade to take charge of the crossing area once it has established its hasty defense at the river. Then the engineer who was responsible for the crossing area becomes the CAE. The brigade XO coordinates with the DMAIN, which retains centralized control of the crossing until only the defending brigade's units remain to cross in that area. The crossing area is used until the commander directs the bridges to be destroyed or removed. At that time, the crossing area ceases to exist.

Turnover of the sites from the CAC to the defending battalion commanders is by mutual agreement or when directed by the brigade commander. Simultaneous handoff between or within defensive sectors is not essential. Depending on the tactical situation, the division commander may not allow crossing equipment to remain in place, even though the defending brigade commander desires its retention. Normally, the CAC retains control of the crossing means until delaying forces cross the river. He then orders the removal of the tactical bridging assets. Control of the remaining fixed bridges then passes to the defending commanders. They are responsible for their defense and ultimate destruction, as discussed in *Chapter 6*.

CHAPTER 4
Planning

GENERAL

Units plan river crossings the same as any tactical operation, with one major difference. Force allocation against enemy units has an added dimension of time. Friendly forces can only arrive on the battlefield at the rate at which they can be brought across the river. This rate changes at different times throughout the operation. This chapter outlines the detailed planning necessary because of this difference.

The corps allocates support elements to the division and provides terrain and enemy analyses. It assigns mission objectives to the division. For operations where the corps is crossing the river, it may assign the bridgehead line.

The division assigns mission objectives to the brigades and specifies the bridgehead line. It may assign bridgehead objectives to the brigades. The division allocates maneuver and maneuver-support forces to the brigades and develops coordination measures, such as movement schedules, that apply to more than one brigade. The division also provides terrain and enemy analyses to the brigades.

The senior corps engineer headquarters, allocated to the division for the crossing, assists the division engineer section with detailed crossing plans. The lead brigade develops the tactical plans that it will execute. It develops the crossing objectives in order to attain its mission objective.

The headquarters of the corps engineer battalion, assigned to support each brigade crossing, develops the detailed crossing plan. The battalions develop the tactical plan necessary to seize assigned objectives.

The actual planning process for a river crossing is the same as for any tactical operation. Differences occur primarily because of the complexity of crossing a river (which makes extensive calculation necessary) and the need to balance tactics with crossing rates.

Planners do crossing calculations twice. Crossing calculations are critical to COA evaluation. They are required to ensure that force buildup supports the COA. For initial planning, simple calculations and rules of thumb are used to produce quick force-buildup information. Once a commander selects a specific COA, planners make detailed crossing calculations to produce the crossing plan.

THE PLANNING PROCESS

The staff planning process produces a best possible solution to accomplish the unit's mission. This chapter discusses those parts of planning that are necessary for a river crossing. It does not attempt to discuss the larger planning process necessary for full mission accomplishment.

In the following paragraphs, the planning process is described in steps and by echelons. The shadowed text in the tables shows the step in the planning process being discussed, with the battle staff and engineer planning requirements alongside. A detailed discussion that is primarily aimed at the division and brigade echelons follows. In general, the corps identifies the crossing requirement and provides assets, the division conducts a detailed terrain analysis and develops rough crossing plans, and the brigade develops detailed crossing plans.

ANALYZING THE MISSION

The first step is to recognize that a river crossing is necessary (see *Table 4-1*). Once the mission is received, the staff develops and conducts a mission analysis. This is done to—

- Understand the purpose of the mission and the intent of the commander and the commander two levels up.

- Review the area of operations.

- Identify tasks (both specified and implied), assets available, constraints, restraints, and an acceptable level of risk.

A mission analysis is conducted according to *FM 101-5*. Corps planners normally identify river-crossing requirements when assigning missions to the division. The corps plan will then provide river-crossing assets to the division and may specify crossing the river as one of the tasks assigned to the division. If the mission the corps is assigning does not require a division-level river crossing, it may not specify a crossing. The troop list includes necessary crossing assets, however.

Normally, if the corps identifies the requirement for a river crossing, its warning order (WO) includes it. The topographic company supporting the corps provides detailed river data and crossing-area overlays. The topographic company automatically provides necessary topographic data to the division terrain team. See *FM 100-15* for more details on planning at the corps level.

The division discovers that it must cross a river by receiving a specified task in the corps's order or by developing an implied task during mission analysis. The division engineer's section always examines all rivers in the division's area of operations during the mission-analysis process. The

Table 4-1. Step 2 - analyze the mission

Military Decision-Making Process	Actions to be Taken

Receive the mission
 	
Analyze the mission
 	
Develop COAs
 	
Analyze COAs
 	
Compare COAs
 	
Approve a COA
 	
Produce orders
	The battle staff—
	
	• Identifies critical facts and assumptions.
	• Conducts an initial IPB by—
	  - Identifying key terrain affecting the crossing.
	  - Templating enemy river defenses.
	  - Estimating the crossing capability of the area to be crossed, using terrain data and available crossing means.
	  - Calculating force crossing rates for each crossing area, using the troop list.
	  - Templating enemy obstacle systems.
	  - Reviewing available bridging assets.
	• Determines specified, implied, and essential tasks.
	• Recognizes that a river-crossing operation is necessary.
	• Issues a WO.
	• Determines the CCIR as pertaining to the river crossing.

division terrain team maintains a terrain database that includes river data and potential crossing sites for the division's area of operation.

NOTE: Upon identifying a river-crossing task, the division engineer and terrain team immediately determine potential crossing sites.

Intelligence Preparation of the Battlefield

The battle staff, including the staff engineer's help, analyze the existing situation. This analysis includes the enemy, friendly troops, terrain, and time available for the mission. This step is primarily designed to acquire the data necessary for the following planning steps, but some early analysis is necessary to generate critical information. The engineer staff officer must very quickly convert raw terrain data and friendly information into crossing rates. This allows the planners to make intelligent decisions about supportable schemes of maneuver.

As a part of the IPB process, the G2 leads the staff development of a defensive situational template along the entire river that the division must cross. The template focuses attention on possible areas of weakness, counterattack forces, and artillery.

The G2, with the division engineer's help, develops obstacle templates from the line of contact through to division objectives. He provides the templates to the brigade intelligence sections for their planning and analysis. The division engineer provides enemy obstacle information (particularly along the river) to the brigade engineers.

The division provides the brigade staff with templates that it refines and further develops for the enemy force in its area of operation. The S2 develops intelligence requirements and a detailed intelligence-collection plan, with specific emphasis on the far shore.

Reconnaissance teams seek information to fill those requirements. Obstacle templates are verified by active air and ground reconnaissance, as directed in *Chapter 2.*

Friendly Troops

The division engineer coordinates for corps engineer units to cross the force, using the simple rule of thumb that every forward brigade requires two bridges. Insufficient bridging assets limit possible COAs.

The brigade engineer identifies the crossing sites required for the brigade and for each battalion based on the number of vehicles. This calculation is based on simple assumptions. From it, the brigade engineer determines the approximate time necessary to cross the entire brigade (see *Appendix B).* The crossings required are important during COA development. The brigade engineer also determines the amount of bridging available, the number of possible heavy rafts, and the number of assault boats. This information is forwarded to the CAE, who is responsible for the control of all crossing means.

Terrain

The division engineer ensures that adequate information is in the crossing-site database for planning at brigade level. The division terrain team generates crossing-site overlays, site data files, and road and cross-country-movement overlays for the crossing areas.

The division engineer ensures that sufficient assault, raft, and bridge sites are available within each assault-brigade area. Generally, a main attack brigade requires assault sites for two dismounted battalions and at least two raft or bridge sites.

The brigade engineer, coordinating with the CAE, evaluates all potential crossing sites from both technical and tactical considerations, including—

- Entry- and exit-road networks.

- Cross-country movement.

- The width, velocity, and depth of the river.

- The conditions of the bank.

- The vegetation along the shore.

- The obstacles in or along the river.

- Possible attack positions and routes to the river.

- Possible call-forward areas.

The brigade engineer, coordinating with the CAE, then analyzes each site to arrive at a rough crossing-rate capability and the effort necessary to open the site. Operational planners use this information to develop possible COAs.

The division engineer, coordinating with the CFC, ensures that the crossing requirements of the lead brigades and breakout force are adequately resourced to satisfy each COA.

The BMAIN evaluates the terrain along the river in terms of OCOKA. The intent is to understand the terrain along the river so that potential COAs can be devised with crossing objectives. The operations planners combine this knowledge with the crossing-site comparisons and enemy templates to develop possible COAs.

DEVELOPING COAs

The G3, along with key members of the battle staff, sketches out possible COAs to accomplish the mission of the division (see *Table 4-2*). COAs must include—

- Assigned crossing areas for each brigade.

- Brigade boundaries that include terrain which is necessary to defend the bridgehead against enemy counterattacks.

Looking two levels down, the division staff plans an assault-crossing site for each anticipated assault battalion in a brigade's area. A brigade should also have two bridging or rafting sites within its boundaries.

The S3 looks closely at the avenues leading to brigade mission objectives, particularly at crossing sites feeding the avenues. Developing practicable COAs is normally an iterative process. The division staff first develops a scheme of maneuver to take the final objective, then verifies that the force

Table 4-2. Step 3 - develop COAs

Military Decision-Making Process	Actions to be Taken
Receive the mission Analyze the mission Develop COAs Analyze COAs	The battle staff— • Sketches out, with the commander's assistance, several COAs to develop. • Develops the scheme of maneuver, fire plan, and support plan for each COA, considering crossing capability and the order of crossing.
Compare COAs Approve a COA Produce orders	The engineer selects sites, determines rafting and bridging configurations and bank-preparation requirements, and task-organizes the engineers for each COA.

build-up rate across the river is adequate for the scheme of maneuver. If so, the S3 expands the COAs to include the tactics required for the crossing.

The tactics required for the crossing are based on enemy defenses near the crossing sites, enemy reaction forces and earliest employment times, and crossing rates at each site. The COAs must include exit-bank, intermediate, and bridgehead objectives.

The S3, working with the brigade engineer and CAE, develops the control measures, crossing graphics, and crossing time line for each COA (see *Figure 4-1*).

ANALYZING COAs

The staff at both the division and brigade war-game each COA against likely enemy reactions (see *Table 4-3, page 4-6*). They then attempt to counter each enemy response.

The engineer—

- War-games against other variables outside his control, such as terrain difficulties and crossing-equipment losses.

- Considers what will happen—

  - If it takes longer to open a crossing site.

  - If damage slows progress over entrance and exit routes.

  - If the conditions of the river change.

- Considers what will happen if enemy action shuts down a crossing site or forces its relocation.

- Must consider the consequences of equipment failure or loss to enemy action.

- Evaluates the most likely of these against all COAs and develops, within his means, necessary counters to include alternate sites and routes.

COMPARING COAs

The division staff examines each COA against both the immediate and follow-on missions (see *Table 4-4, page 4-6*). The division is particularly concerned with the movement of reserve and support forces and compares COAs against these requirements.

The brigade staff considers the ability of each COA to handle enemy responses, support follow-on missions, provide brigade flexibility, and allow for crossing redundancy.

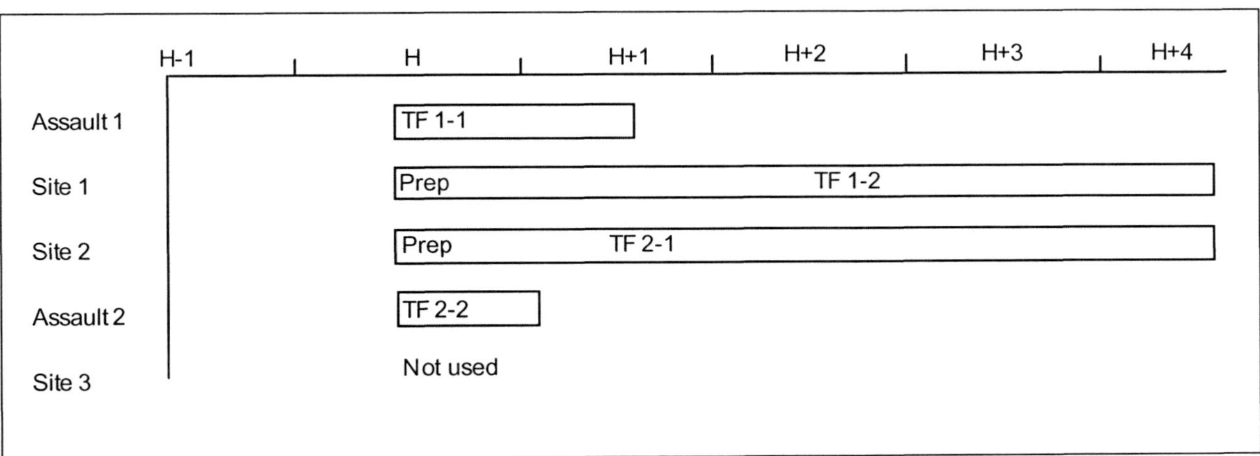

Figure 4-1. Crossing time line

Table 4-3. Step 4 - analyze COAs

Military Decision-Making Process	Actions to be Taken
Receive the mission Analyze the mission Develop COAs Analyze COAs Compare COAs Approve a COA Produce orders	The battle staff war-games each COA against possible enemy responses. The engineer war-games each COA against terrain changes and equipment loss.

PRODUCING ORDERS

The battle staff converts the selected COA into a plan with sufficient detail for synchronized execution (see *Table 4-5*). The staff engineer conducts an extensive analysis to develop a unit-by-unit crossing plan and movement schedule in conjunction with the G3, G4, and DTO. From this analysis, he develops the crossing-capability chart (see *Appendix B*) and the crossing overlay (see *Figure 3-6, page 3-11)*. These are his primary execution tools. The staff engineer develops the crossing-synchronization matrix as a primary execution tool for the S3 (see *Appendix B*). He also helps the traffic-control cell work out the traffic-circulation plan.

While detailed planning is underway, the CAE initiates far-shore and nearshore reconnaissance to develop sufficient detail for battalion-level planning. He converts this planning into a detailed engineer task list and develops an engineer execution matrix to synchronize it (see *Appendix B*).

Table 4-4. Step 5 - compare COAs

Military Decision-Making Process	Actions to be Taken
Receive the mission Analyze the mission Develop COAs Analyze COAs Compare COAs Approve a COA Produce orders	The battle staff— • Compares and evaluates the advantages and disadvantages of the COAs. • Recommends one COA to the commander. The commander selects a COA and issues a FRAGO.

Table 4-5. Step 6 - produce orders

Military Decision-Making Process	Actions to be Taken
Receive the mission	The battle staff converts the selected COA into an executable plan.
Analyze the mission	The engineer develops a detailed crossing plan.
Develop COAs	
Analyze COAs	
Compare COAs	
Approve a COA	
Produce orders	

CHAPTER 5
Division Deliberate River Crossing

GENERAL

A division deliberate river crossing is an operation conducted as part of an offensive operation. The intent of a deliberate river crossing is to quickly cross a river and rapidly secure the bridgehead line. It is meticulously planned and coordinated with all concerned elements. A deliberate river crossing requires thorough reconnaissance and extensive evaluation of all intelligence. It requires detailed planning and preparation, centralized control, and extensive rehearsals.

A deliberate river crossing is costly in terms of manpower, equipment, and time. It is generally conducted against a well-organized defense when a hasty river crossing is not possible or when one has failed. A deliberate river crossing requires the concentration of combat power on a narrow front, capitalizing on the element of surprise. The phases, echelons, organizations, and C^2 of a division deliberate river crossing are discussed in detail in this chapter.

PHASES OF A DELIBERATE RIVER CROSSING

An offensive deliberate river-crossing operation has four phases. They are distinct phases for planning, but there is no pause between them in execution. The phases are as follows:

- *Advance to the river (Phase I).* The first phase is the attack to seize the near-shore objective.

- *Assault across the river (Phase II).* The second phase involves units assaulting across the river to seize the far-shore objective, eliminating direct fire on the crossing sites.

- *Advance from the exit bank (Phase III).* The third phase is the attack to secure exit-bank and intermediate objectives that eliminate direct and observed indirect fires into the crossing area.

- *Secure the bridgehead line (Phase IV).* The final phase involves units that secure bridgehead objectives to protect the bridgehead against a counterattack. This gains additional time and space for the buildup of forces for the attack out of the bridgehead.

These phases are followed immediately by an attack out of the bridgehead by follow-on forces to defeat enemy forces at subsequent or final objectives. *Figure 5-1, page 5-2*, relates the crossing phases to the objectives described in this chapter.

THE RIVER CROSSING

The following section describes a deliberate river-crossing operation from the division's and brigade's perspectives. It details the actions that are required in deep, close, and rear operations by phase (see *Figure 5-2, page 5-3*).

A division is normally the smallest organization that can conduct a deliberate river-crossing operation. It is usually an implied task in a larger mission given by the corps. The river crossing is not the objective but is part of the scheme of maneuver and

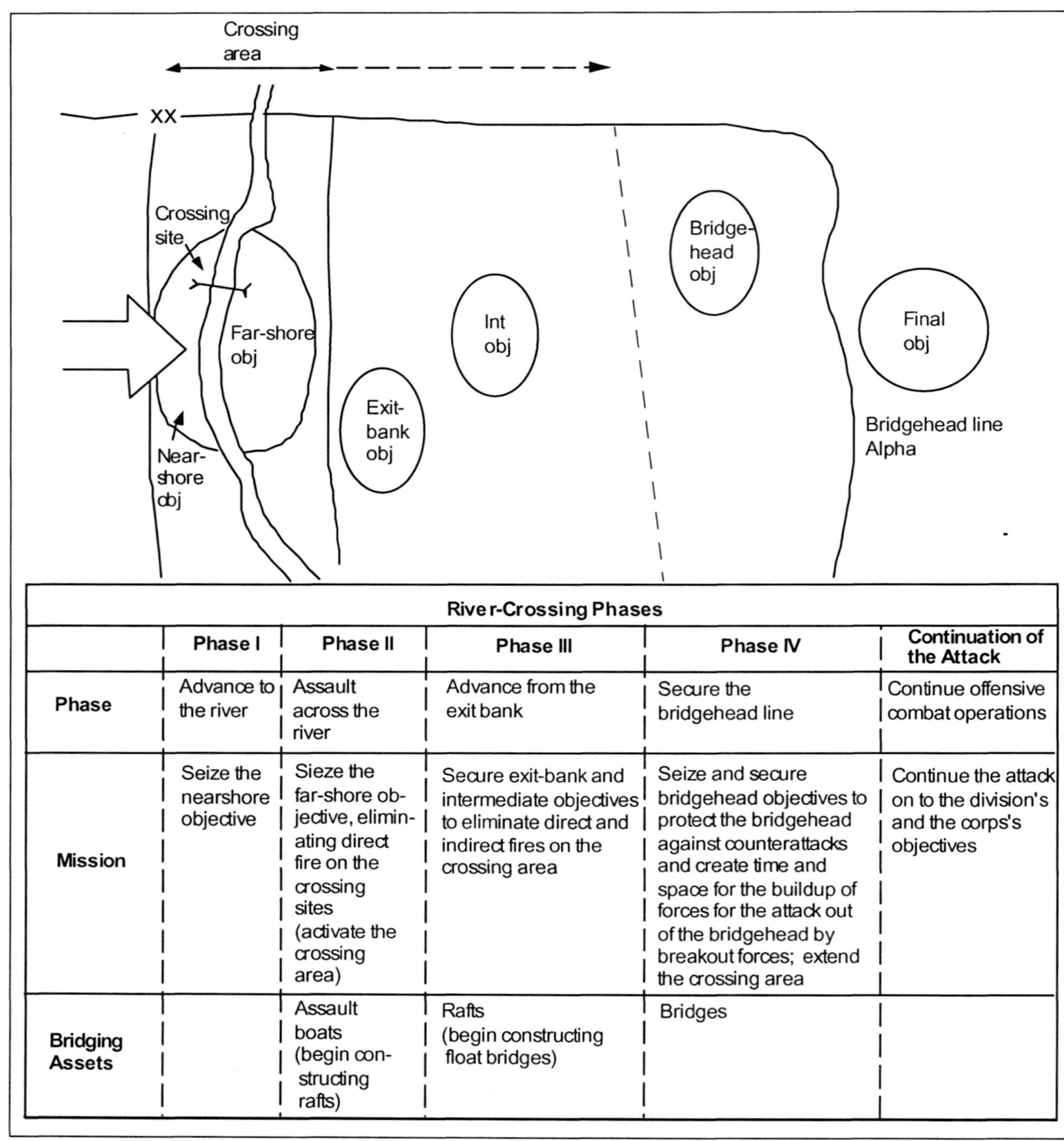

River-Crossing Phases					
	Phase I	**Phase II**	**Phase III**	**Phase IV**	**Continuation of the Attack**
Phase	Advance to the river	Assault across the river	Advance from the exit bank	Secure the bridgehead line	Continue offensive combat operations
Mission	Seize the nearshore objective	Sieze the far-shore objective, eliminating direct fire on the crossing sites (activate the crossing area)	Secure exit-bank and intermediate objectives to eliminate direct and indirect fires on the crossing area	Seize and secure bridgehead objectives to protect the bridgehead against counterattacks and create time and space for the buildup of forces for the attack out of the bridgehead by breakout forces; extend the crossing area	Continue the attack on to the division's and the corps's objectives
Bridging Assets		Assault boats (begin constructing rafts)	Rafts (begin constructing float bridges)	Bridges	

Figure 5-1. Deliberate river crossing

overall offensive action against the enemy. The enemy will normally use the river as a tactical obstacle system to slow and gain positional advantage against the division's advance. The intent of the division is to maintain its momentum through the crossing.

Mission, enemy, terrain, troops, and time available (METT-T) dictate the force allocation required during each phase of the operation. Aside from the normal planning, detailed march tables are required for the rapid passage of units through the crossing area into the bridgehead. Detailed plans are disseminated before the execution to ensure an uninterrupted operation. River-crossing operations normally restrict movement to four to six routes. This requires disciplined and controlled movement to ensure that combat power builds in the bridgehead faster than the enemy's ability to react.

An integral part of the river-crossing operation is the deception plan. The corps will plan, resource, and control all of the requirements to execute a believable deception so that the enemy does not know where the division will conduct the deliberate river-crossing operation.

To conduct the deliberate river crossing, the division requires augmentation from the corps. The corps must provide bridge companies that are in direct support to the division for the river-crossing operation in addition to other combat engineers that are required to operate assault boats, provide C^2, and so forth. An assault float bridge (AFB) company must have an engineer group or ad hoc battalion staff that can support the deliberate river-crossing operation and can remain in

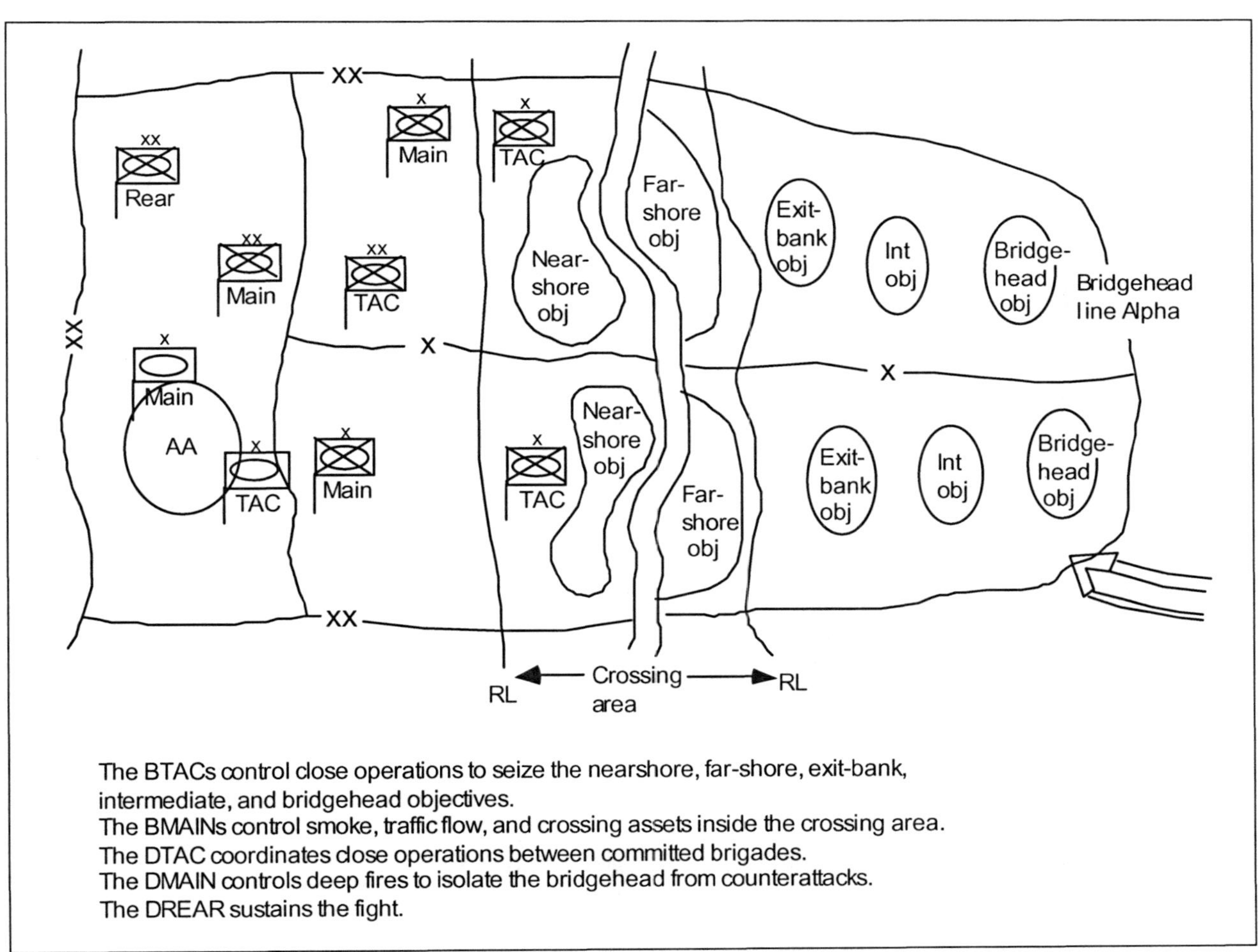

The BTACs control close operations to seize the nearshore, far-shore, exit-bank, intermediate, and bridgehead objectives.
The BMAINs control smoke, traffic flow, and crossing assets inside the crossing area.
The DTAC coordinates close operations between committed brigades.
The DMAIN controls deep fires to isolate the bridgehead from counterattacks.
The DREAR sustains the fight.

Figure 5-2. Division and brigade CP functions

place after the division continues the attack to subsequent corps objectives. Engineer groups should include one corps combat engineer battalion and two AFB companies for each lead brigade. The corps normally provides a corps engineer light diving team to—

- Conduct nearshore and far-shore reconnaissance.

- Perform bottom-composition surveys.

- Neutralize underwater obstacles.

- Construct underwater bridge structures, obstacles, and floating barriers.

- Perform in-water repair to bridges and watercraft.

- Recover sunken equipment.

- Search for and recover casualties.

Additionally, the corps normally provides a corps MP company to assist the division in regulating the traffic and conducting route security in the crossing area. The corps also allocates additional smoke units to assist the division chemical company in obscuring the river-crossing area. Finally, the corps will provide short-range air-defense (SHORAD), high-to-medium-altitude air-defense (HIMAD), and air-defense artillery (ADA) support to protect the bridgehead from air interdiction.

ADVANCE TO THE RIVER (PHASE I)

Once the division has planned the operation, the first phase is initiated (see *Chapter 4*). The division will attack to seize nearshore terrain that includes favorable crossing sites and road networks. Normally, the division advances with two brigades abreast and a reserve brigade trailing. The cavalry squadron can provide a forward or flank screen (see *Figure 5-3*). The DTAC controls the efforts of the lead brigades (see *FM 71-100*).

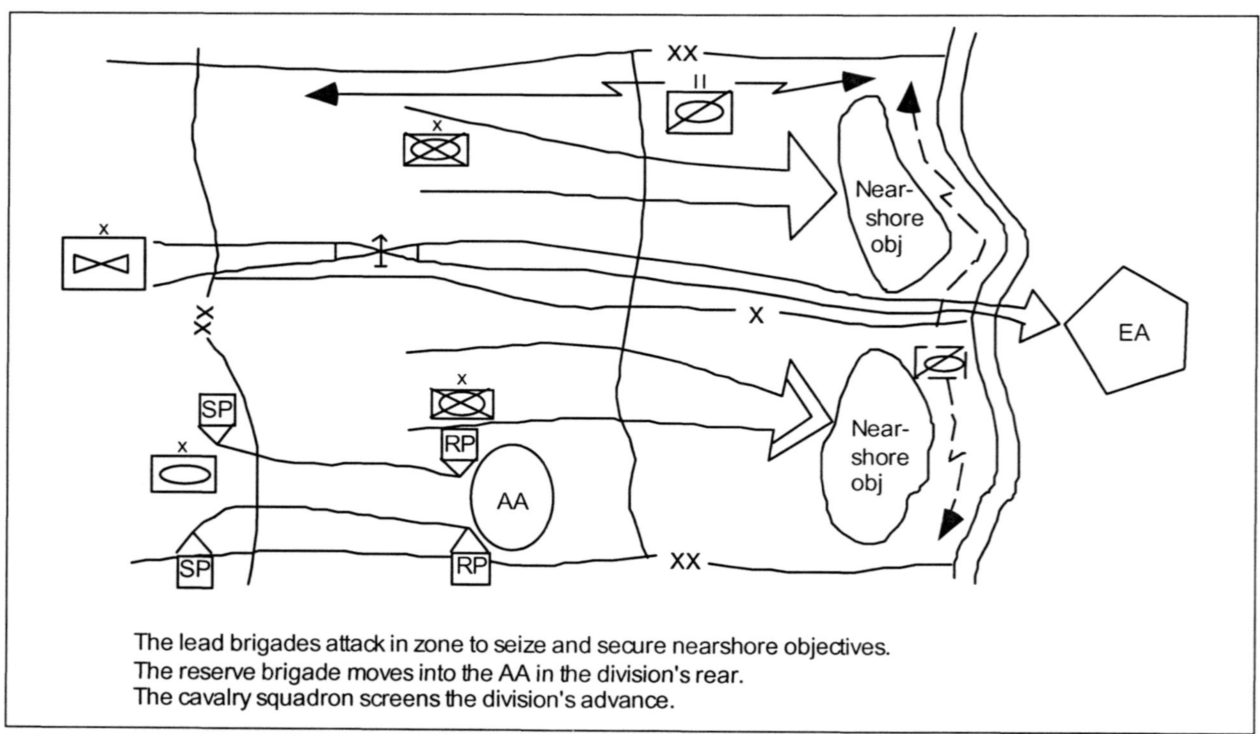

The lead brigades attack in zone to seize and secure nearshore objectives.
The reserve brigade moves into the AA in the division's rear.
The cavalry squadron screens the division's advance.

Figure 5-3. Advance to the river (division focus)

Well before the division reaches the river, the cavalry squadron moves ahead of the main body to conduct a reconnaissance of the near shore and predetermined crossing sites. Engineer reconnaissance teams may need to be allocated to the division cavalry squadron to assist in the reconnaissance of crossing sites. If the tactical situation prohibits the cavalry squadron from moving to reconnoiter the crossing sites, one or both of the lead brigades must conduct the reconnaissance. As the division arrives at the river, the lead brigades establish security on the near shore. The lead brigades develop hasty defensive positions to protect the crossing area and cover the crossing sites with direct and indirect fires.

During the advance to the river, the DMAIN coordinates counterfires, close air support (CAS), and support of the division aviation brigade against deep targets. By effectively using these assets, the DMAIN fights the deep battle and isolates the bridgehead.

The DREAR sustains the division's advance. It ensures that key classes of supplies are pre-positioned forward. Priority is shifted to the maintenance of the bridging assets and those of the units supporting the crossing area.

The BTAC controls the fight of the TFs within its brigade. The brigade travels in a formation that is METT-T driven. The brigade seizes objectives that secure the near shore (see *Figure 5-4*).

Each BMAIN is the crossing-area headquarters. The crossing area is bounded by RLs

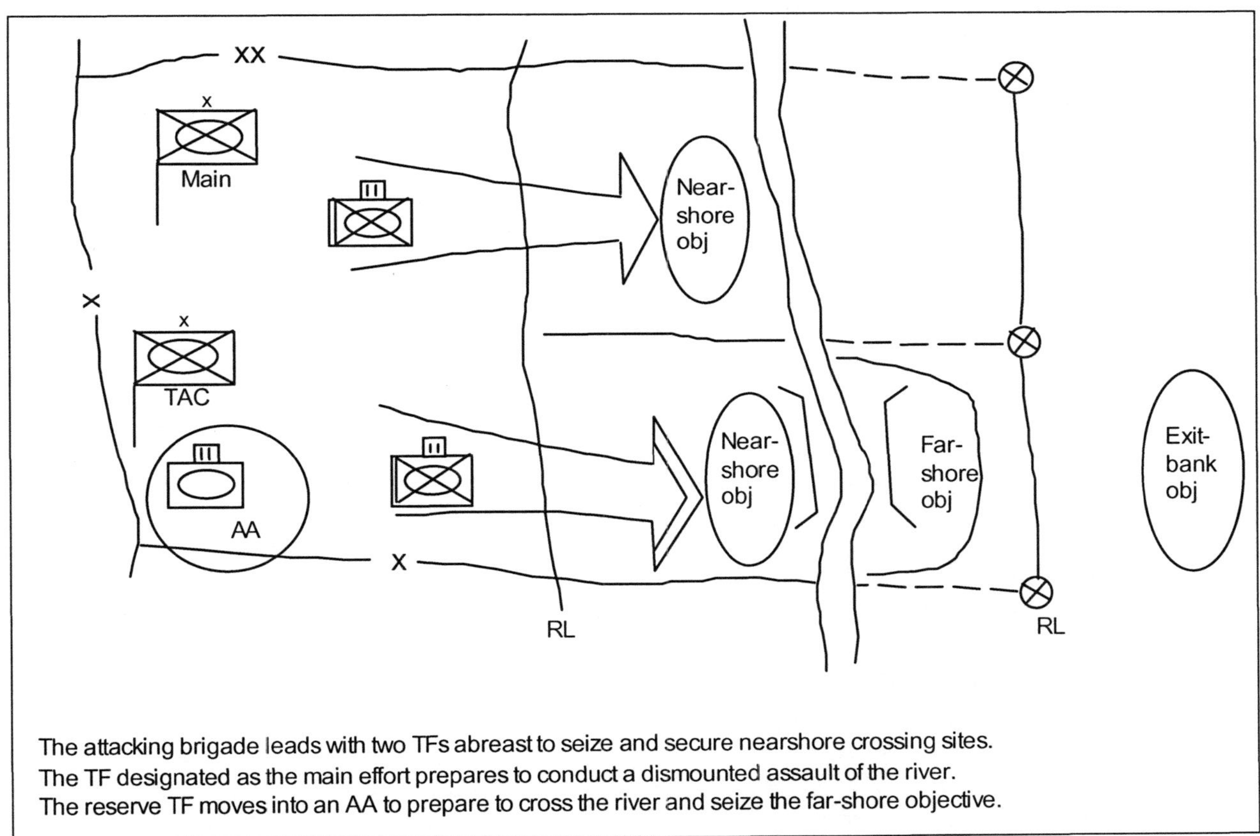

The attacking brigade leads with two TFs abreast to seize and secure nearshore crossing sites.
The TF designated as the main effort prepares to conduct a dismounted assault of the river.
The reserve TF moves into an AA to prepare to cross the river and seize the far-shore objective.

Figure 5-4. Advance to the river (brigade focus)

on the friendly and enemy sides of the river. The RL on the friendly side is usually set 2 to 3 kilometers from the exit bank, out of the range of enemy direct-fire weapons. The RL on the enemy side delineates an area large enough for forces to occupy battalion-sized attack positions. The BMAIN is responsible for controlling units that provide the crossing means, traffic management, and obscuration. Normally, corps assets are task-organized by the division in direct support of the forward brigades to perform these functions. The BMAIN controls these assets. Once the brigade has secured the near shore, MP and engineers mark routes from the staging area to the crossing sites; lay out staging, holding, and call-forward areas; and set up ERPs and TCPs.

Once the near shore is secured, the DTAC becomes the crossing-force headquarters responsible for coordinating the close operations of the committed brigades within the bridgehead and crossing area. The bridgehead is the area on the far shore that is required to provide space and time for the buildup of combat power to continue offensive combat operations. The crossing area is the area, bounded on either side of the river by RLs, in which units move on predetermined routes and use the time tables that are specified in the division's order.

The DTAC coordinates the efforts of the lead brigades as they prepare to assault across the river.

ASSAULT ACROSS THE RIVER
(PHASE II)

The DMAIN continues to control deep-fire assets to isolate the bridgehead. As units advance, deep fires shift to subsequent targets. The division coordinates with the corps for SHORAD and HIMAD coverage to protect the bridgehead from enemy air interdiction. The corps normally provides

Patriot and Hawk support. The division AD battalion provides local AD coverage. The river creates lucrative targets at relatively fixed locations that are easily targeted by enemy air. Therefore, approaches; holding, staging, and call-forward areas; and crossing sites along the river are the highest priority for AD during the crossing. AD units occupy positions to engage aircraft with massed fires before the aircraft can reach weapons release points (RPs).

The DTAC coordinates the actions of the brigades conducting the assault across the river (see *Figure 5-5*). The crossing sites are chosen because of available concealment, a good route system, and sufficient space for AAs on the near shore. These sites also have defensible terrain on the far shore of the river to provide a secure base for continuing the operation.

The DREAR begins to push packages of Class IV and V supplies to support the hasty defense to secure the bridgehead line.

The BTACs control their own respective assault-crossing elements, which normally consist of dismounted infantry. A corps combat engineer company, operating assault boats (RB15s) from the corps bridge companies, transports the dismounted soldiers of the assault force to the far shore. The dismounted element crosses the river and secures terrain for the reinforcing armored vehicles. The assault across the river can also be an air-assault operation. The dismounted assault forces are supported by the tanks and infantry fighting vehicles from their TF and by other combat units in support-by-fire positions. Heavy rafts are prepared to transport tanks and infantry fighting vehicles to the far shore for reinforcing the dismounted infantry. M9 ACEs/dozers are

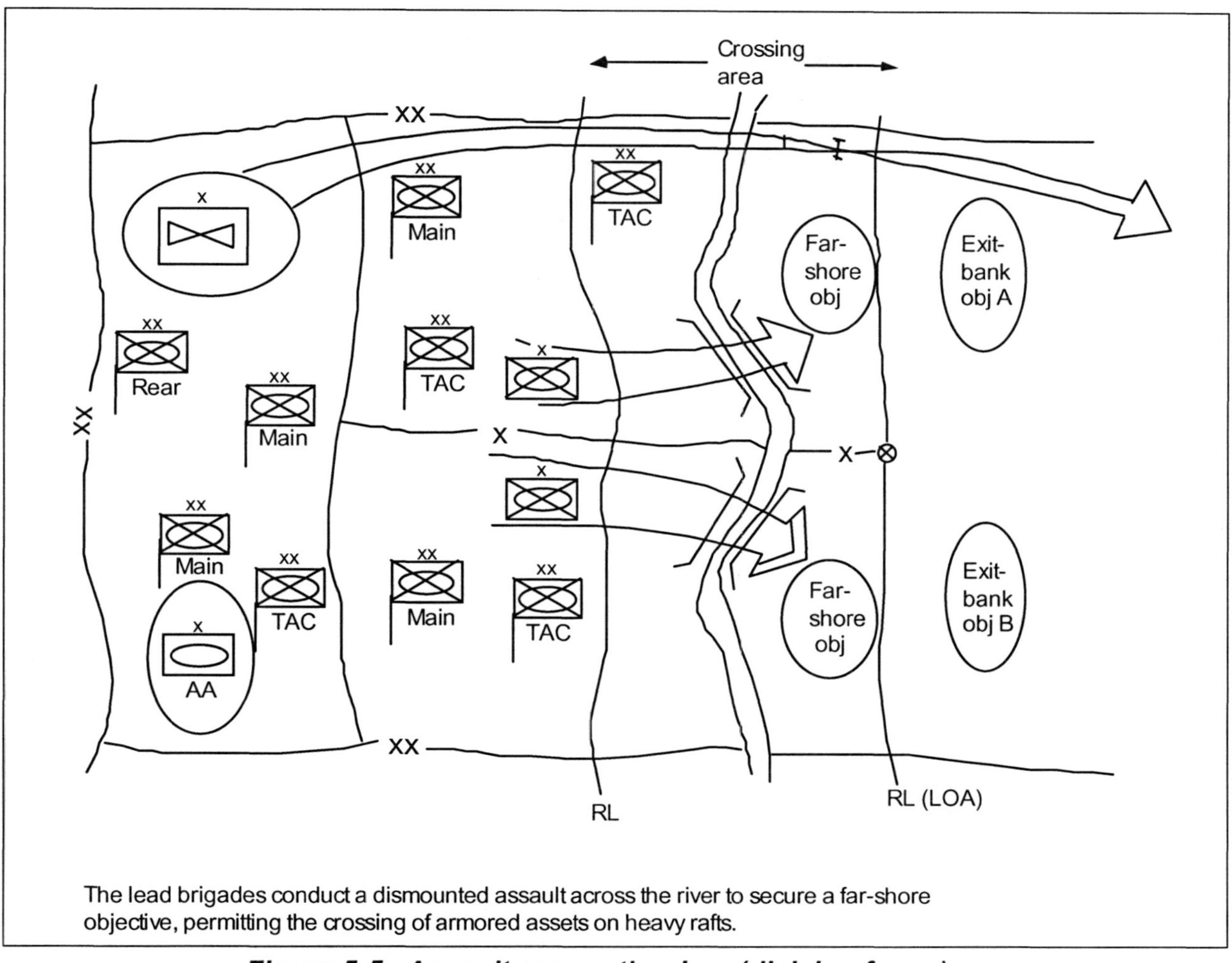

Figure 5-5. Assault across the river (division focus)

transported to prepare the far-shore exit sites. Rapid reinforcement of dismounted assault forces with armored vehicles may be so critical, based on the METT-T, that it justifies using any expedient method to get the first few armored vehicles across. This includes winching, towing, or pushing the first ones across normally unsuitable places while engineers improve entry and exit points for the rest.

Each BMAIN controls smoke to obscure crossing sites on the river. When employed, the smoke blanket covers several kilometers of the river and river approaches to conceal the actual crossing locations, but not as to obscure the bridge crewmens' vision. The crossing-area headquarters uses smoke generators, smoke pots, and smoke munitions from the division and corps. The BMAIN controls the use of MP and corps engineer units to establish nearshore waiting areas, mark routes to the crossing sites, and begin constructing heavy rafts and/or bridges.

The intent of this phase is to rapidly place combat power on the far shore to eliminate the enemy's direct fire onto the crossing sites and secure terrain for attack positions.

Brigades normally establish limits of advance (LOAs) and fire-support coordination lines (FSCLs) for the dismounted TFs conducting the assault. These lines establish an LOA that encompasses the far-shore objective. Enemy indirect fire into the crossing area will probably continue; however, each crossing site within the crossing area must be isolated from direct fire to enable the construction and operation of rafts. These rafts will then be used to transport armored vehicles for rapid reinforcement of the dismounted infantry TF. Within the crossing area, secured attack positions allow units to form into combat formations before continuing the attack.

Commanders may consider immediate construction of a bridge during this phase without ever conducting rafting operations. The advantage is that combat power can be massed on the far shore at a much faster rate. The risk that the commander takes in making this decision is that a large amount of bridging assets is exposed to enemy fire before the elimination of enemy indirect fires on the crossing area.

ADVANCE FROM THE EXIT BANK
(PHASE III)

The division continues its attempt to secure the bridgehead line by attacking to seize and secure exit-bank and intermediate objectives. The intent is to eliminate direct and observed indirect fires from the crossing area (see *Figure 5-6*).

The division commander selects exit-bank and intermediate objectives based on METT-T. The river splits the attacking force, limiting massed direct fires beyond the exit bank. Therefore, these objectives are usually smaller and not as far from the attack positions as the objectives used in other offensive operations.

Once the exit banks are secured, the division cavalry squadron crosses either by swimming or rafting their cavalry fighting vehicles. They then conduct normal screening operations for the division as the armored reinforcements are crossing the river and preparing to advance from the exit bank.

The DTAC controls the coordinated attack of the lead brigades and the cavalry squadron to seize exit-bank and intermediate objectives. The DMAIN controls deep fires that aviation, artillery, and CAS provide to block enemy counterattacks into the bridgehead as requested by the DTAC.

The DREAR prepares to push packages of Class III and V supplies that will support the attack out of the bridgehead. They also begin to push Class IV and V supplies for the hasty defense during the last phase of the river-crossing operation.

The BMAINs control the movement of their follow-on TFs from the staging areas across the river to their attack positions on the far shore. They control the upgrade of crossing sites from assaults boats (RB15s) to heavy rafts and/or bridging to ensure that the force buildup can support the advance from the exit bank to intermediate objectives. MP and, if available, corps combat engineers assist in movement control through the crossing area.

During this phase, limited two-way traffic begins to return disabled equipment and casualties.

The BTAC controls the movement out of the attack positions to exit-bank and intermediate objectives. Exit-bank objectives are those positions that, when seized, eliminate the enemy's ability to use direct-fire weapons on the crossing area. Intermediate objectives are those positions from which the enemy can provide observation for indirect-fire

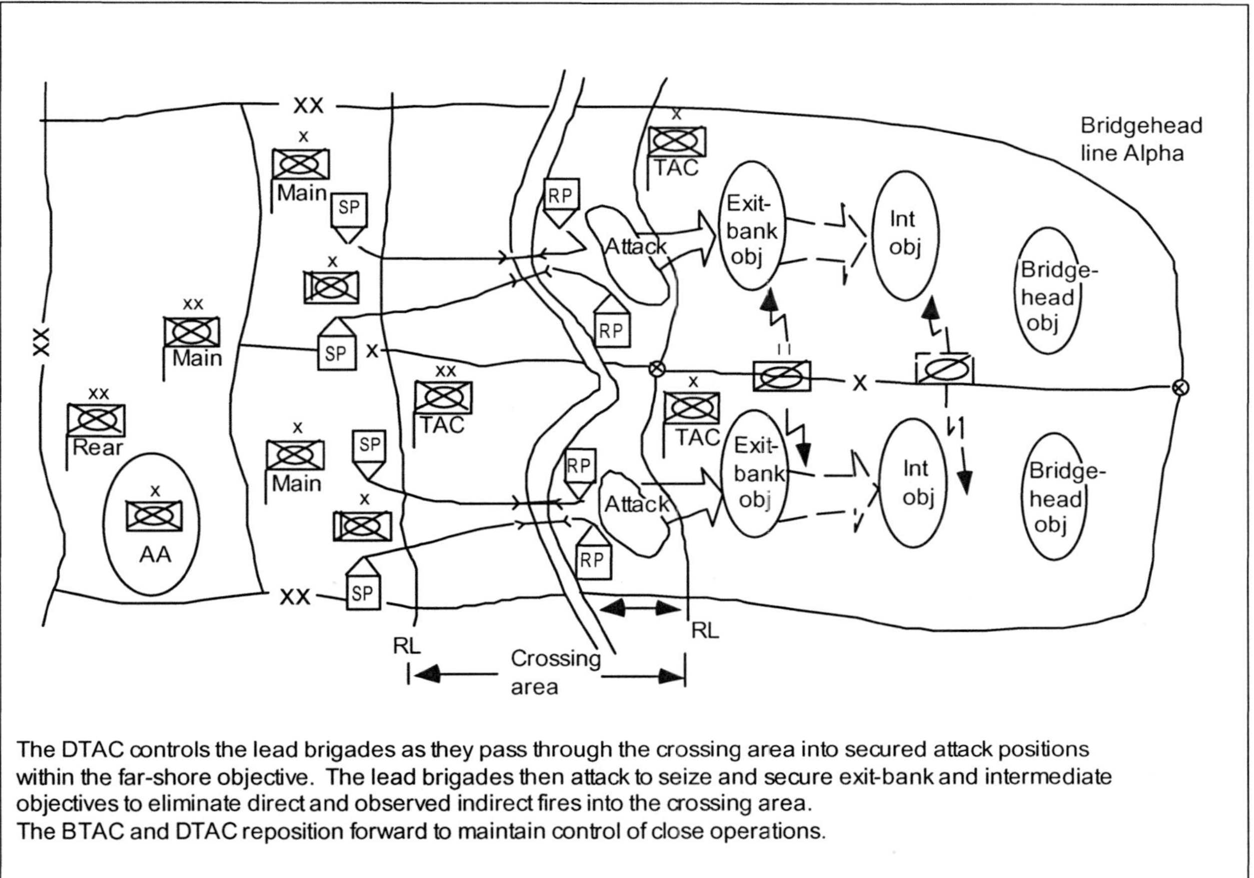

The DTAC controls the lead brigades as they pass through the crossing area into secured attack positions within the far-shore objective. The lead brigades then attack to seize and secure exit-bank and intermediate objectives to eliminate direct and observed indirect fires into the crossing area.
The BTAC and DTAC reposition forward to maintain control of close operations.

Figure 5-6. Advance from the exit bank (division focus)

weapons. This enables the expansion of SHORAD coverage, allowing more time to engage aircraft in air avenues of approach on the far shore (see *Figure 5-7, page 5-10*).

The TF that conducted the dismounted assault across the river continues to cross armored vehicles and remount their dismounted soldiers in preparation for continued offensive operations.

The brigade commanders establish the order of raft loads based on the division's crossing priorities. Bridge companies run heavy raft sites and begin to construct ribbon bridges. MP mark routes and control holding areas on the far shore to ensure rapid transit within the crossing area.

SECURE THE BRIDGEHEAD LINE (PHASE IV)

The bridgehead must be defendable and large enough to accommodate forces that will break out to continue offensive combat operations. The lead brigades attack to secure the final objectives within the bridgehead to prevent the enemy from successfully counterattacking against forces within the bridgehead line by rapidly building enough combat power to establish a hasty defense in the sector. The cavalry squadron conducts a screen mission. The

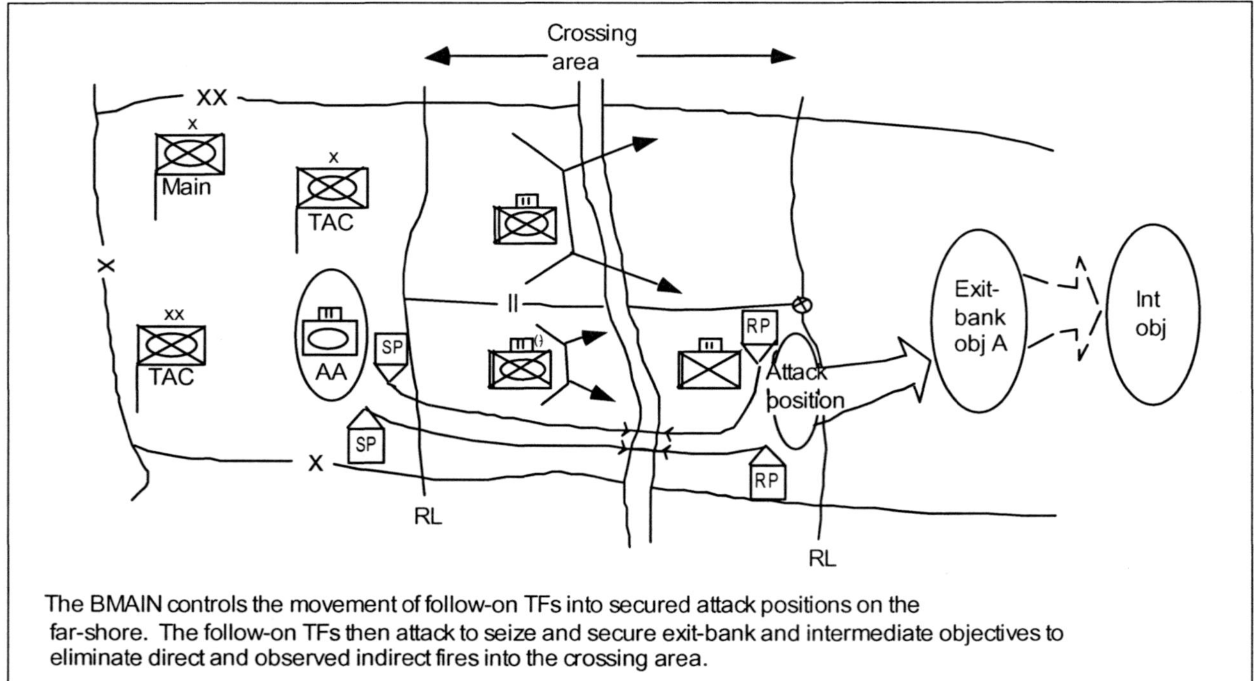

The BMAIN controls the movement of follow-on TFs into secured attack positions on the far-shore. The follow-on TFs then attack to seize and secure exit-bank and intermediate objectives to eliminate direct and observed indirect fires into the crossing area.

Figure 5-7. Advance from the exit bank (brigade focus)

lead brigades maintain continuous far-shore security to prevent bypassed enemy elements from infiltrating back to the river and disrupting activities at the crossing sites (see *Figure 5-8*).

The DTAC controls the lead brigades and the cavalry squadron as they secure the bridgehead objectives (see *Figure 5-8*) and prepare to move the reserve brigade or other corps forces (breakout forces) into attack positions within the bridgehead. Once the bridgehead objectives are secured, the lead brigades establish a hasty defense in sector.

The DREAR begins to push forward Class III and V supplies that are needed for the attack out of the bridgehead.

The BMAIN continues to upgrade and monitor the crossing sites and control the movement of forces through the crossing area. The far-bank RL, defining the crossing area, is moved just past the inter-

mediate objectives (see *Figure 5-8*) to provide space for the breakout forces. Once the bridgehead line is secure, the DTAC controls the movement of the breakout forces through the crossing area to attack positions within the bridgehead. During this phase, specific bridges and/or rafts are designed for full-time return traffic. This ensures that resupply and the evacuation of wounded soldiers and disabled equipment occur.

The DMAIN controls the aviation, artillery, and available CAS sorties to screen the flanks and interdict enemy counterattacks. Deep operations play a key role in the bridgehead defense by targeting enemy formations as they move to counterattack. They also eliminate effective artillery fire within range of the bridgehead and destroy other enemy artillery forces moving up to the fight.

The lead brigade elements that secure the bridgehead line must control the avenues of approach into the bridgehead and be large

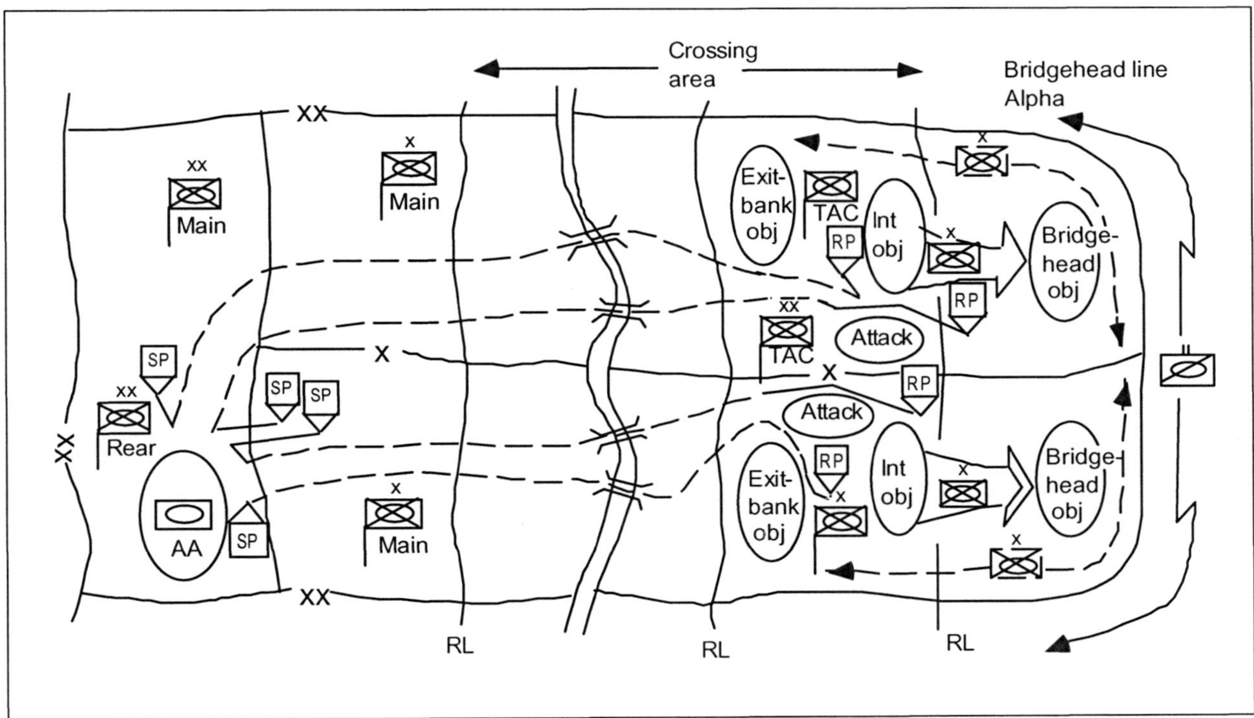

Figure 5-8. Secure the bridgehead line (division focus)

enough to defeat counterattacks. After the bridgehead is secure, the division commander commits the breakout force to attack position within the bridgehead. The bridgehead needs enough space (20 to 30 kilometers deep) to accommodate both the lead brigades and the breakout force with their combat service support (CSS). The bridgehead line must also be deep enough to employ AD systems against hostile aircraft before they reach weapons RPs to attack crossing sites.

CONTINUATION OF THE ATTACK

Once the division has secured the bridgehead, the division river crossing is complete. Crossing-area control will be passed to the DREAR and ultimately to the corps. The breakout force must complete its passage before continuation of offensive operations. The lead brigades must reorganize and prepare to follow the breakout force as the division or corps reserve. Security

forces from the corps must come forward to relieve the lead brigades from their bridgehead security mission.

As the breakout force crosses into attack positions, the DTAC begins to focus on the attack out of the bridgehead. Therefore, the DREAR assumes the role of the crossing-force headquarters. This allows the DTAC to focus completely on the attack out of the bridgehead, which is usually led by the division cavalry squadron.

The DREAR controls the breakout force's movement through the crossing area to the attack positions and two-way traffic facilitating the return of wounded soldiers and disabled equipment. The corps must provide other forces for bridgehead security before the lead brigades reorganize to resume their mission as the division reserve.

CHAPTER 6
Retrograde Operations

GENERAL

The goal of a retrograde river-crossing operation is to cross a river while preserving the integrity of the force. A retrograde operation is an organized movement to the rear or away from the enemy.

This chapter describes only those tactics and techniques used by a division in a retrograde river-crossing operation that are different from those used in an offensive crossing. A retrograde crossing features centralized control at division level. Detailed planning and preparation of engineer assets are a critical consideration within the time available. A retrograde crossing differs from an offensive crossing in several aspects:

- Both banks of the river initially are under friendly control. Accordingly, detailed information concerning the river and the area over which the retrograde is conducted should be readily available to the commander.

- All existing bridges and other crossing sites are available to the retrograde force to expedite the crossing.

- Relative combat power favors the enemy in most cases. Units conducting retrograde operations then must retain a mobility advantage over the enemy.

Deception is always planned and executed to deceive the enemy and to protect the force during a retrograde operation. As a minimum, these plans seek to conceal the extent of the operation and the actual crossing sites. Smoke, electronic deception, and dummy sites reduce the enemy's capability to disrupt the crossing.

The same control measures are used in retrograde operations as in offensive operations. *Figure 6-1, page 6-2,* shows an example. See *Chapter 3* for a discussion of each control measure and a C^2 diagram.

RETROGRADE TYPES

A retrograde operation may be forced by enemy action or by a higher headquarters. A well-planned, well-organized, and aggressively executed retrograde operation provides opportunities for the division to inflict heavy damage on enemy troops and equipment while continuing to maintain its fighting integrity. The three types of retrograde operations are delay, withdrawal, and retirement.

DELAY

Units conduct delays when their strength is insufficient to attack or defend or when they want to maneuver the enemy into an area for a subsequent counterattack. A delay is an operation in which the unit, under enemy pressure, trades space for time by inflicting maximum damage on the enemy without being decisively engaged in combat. Flexible planning allows the units conducting a river crossing to adapt quickly to changes during execution. Some important features of a flexible plan include—

- Multiple approach routes from battle positions to crossing sites.

- Lateral routes between crossing sites.

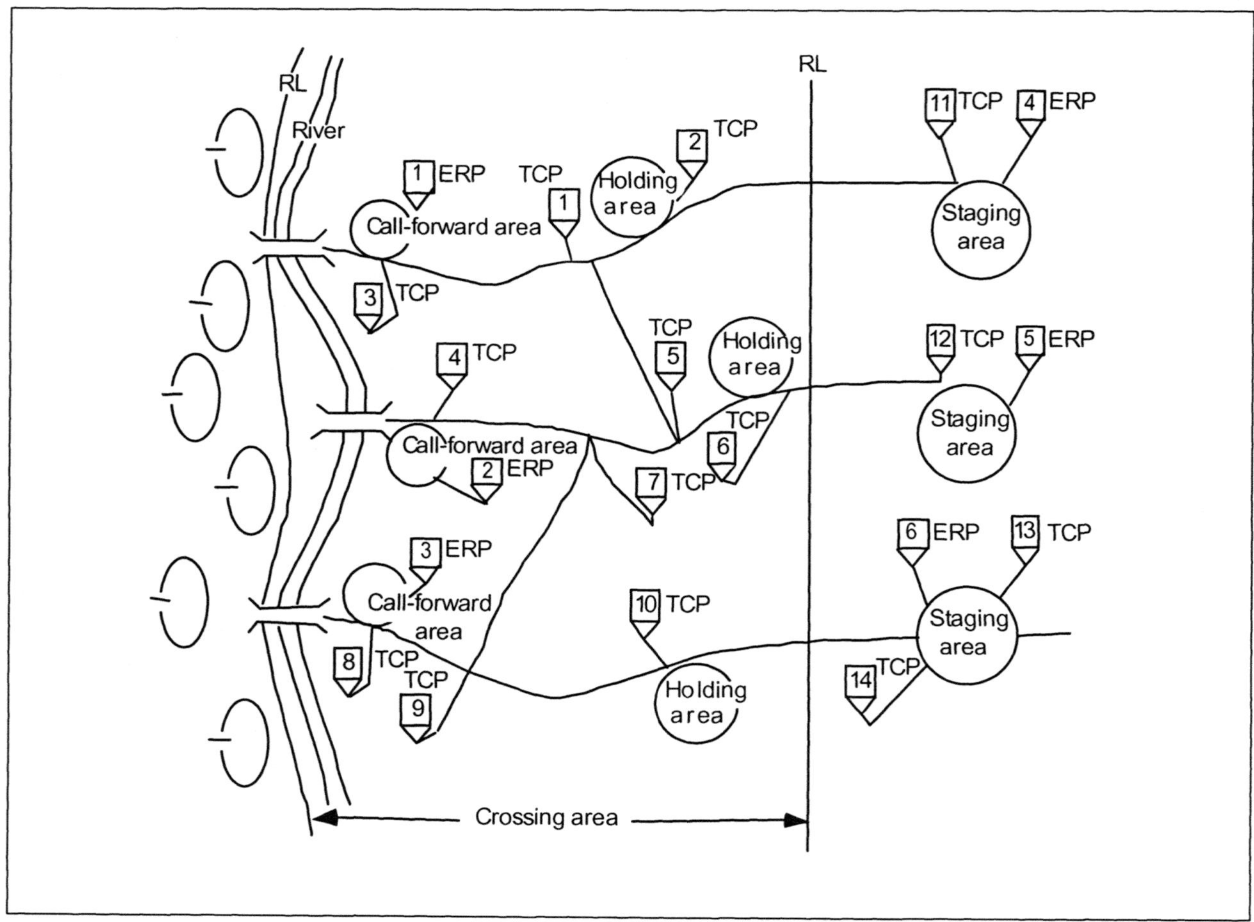

Figure 6-1. Control measures

- Alternate crossing sites if enemy actions close primary sites.

- Crossing equipment held in reserve to replace losses or open alternate sites.

- Preplanned engagement areas (EAs) to block enemy advances.

A delay combined with a retrograde river crossing has the following phases:

- Delay.

- Crossing.

- Defense.

Each phase is separate only in planning; they overlap during execution. Employing military crossing equipment in the retrograde is the reverse of the method used in a deliberate, offensive river-crossing operation. *Figure 6-2* relates the retrograde sequence to the crossing stages.

Delay Phase

The delay phase provides security for the main body and allows the delaying force to gain enough time for the main body to accomplish its mission (cross the river). For this reason, delaying forces take some risk. The delaying force must deceive the enemy and keep it from the river, allowing the

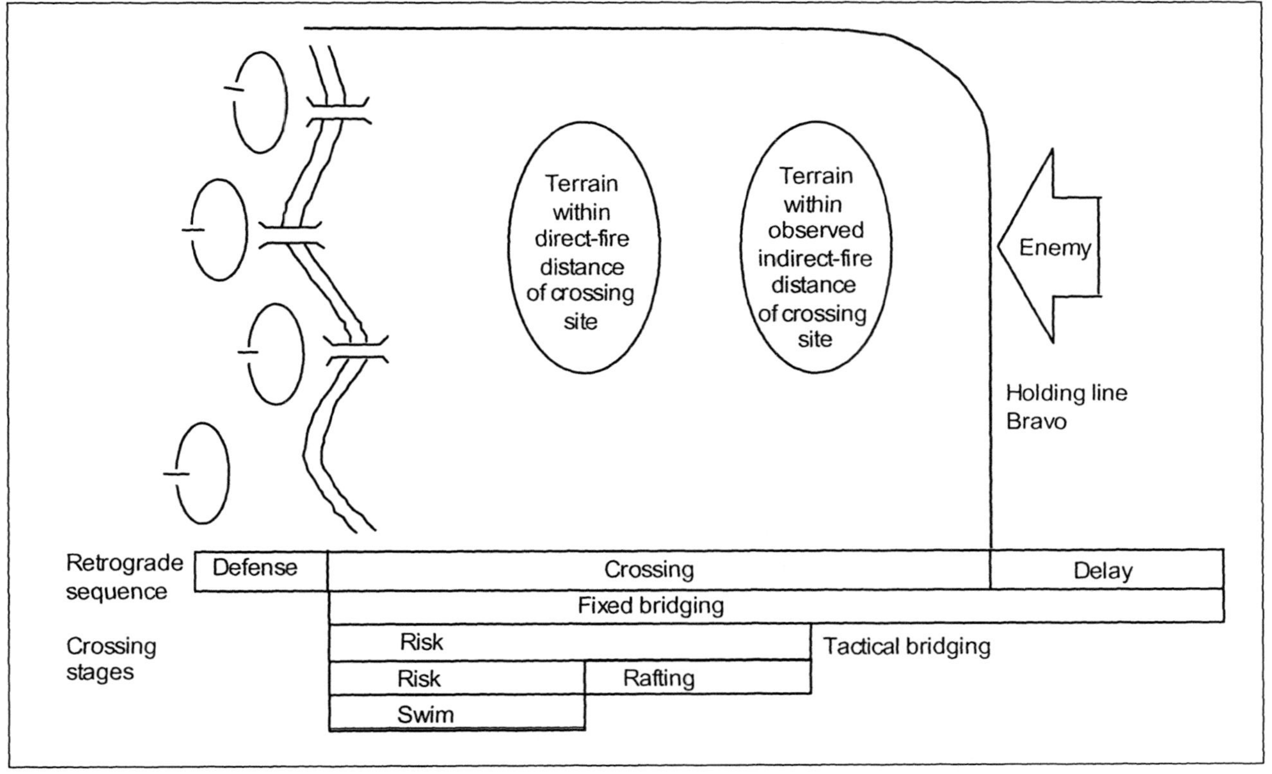

Figure 6-2. Retrograde planning

main body to cross and establish the exit-bank defense.

The division commander establishes a holding line on defensible terrain between the river and the enemy. Its location precludes direct and observed indirect fires in the crossing area.

Forces not assigned tasks in the delay, including those forces with a mission to support crossing areas or establish the defense on the exit bank, execute a planned retirement or withdrawal and cross the river as rapidly as possible. To preclude early enemy detection of the retrograde, the forces follow a movement-control plan that supports the deception plan.

The delay phase continues until the battle is within communications and fire-support range of the exit-bank defense. The delaying

force must be strong enough to hold the enemy until other forces establish the defense. The defending force assumes responsibility for the battle as the delaying force completes a rearward passage of lines through the defending force.

Figure 6-3, page 6-4, shows an example of a retrograde crossing. In this case, the 3rd brigade is the delaying force. It occupies battle positions to the rear of the 1st and 2nd brigades at RL Plum, the initial-delay position (IDP), to help them withdraw. The 3rd brigade delays the enemy forward of the holding line until the rest of the division crosses the river and the 1st and 2nd brigades reestablish the defense along the river.

Crossing Phase

In contrast to normal offensive crossing operations, friendly forces initially control

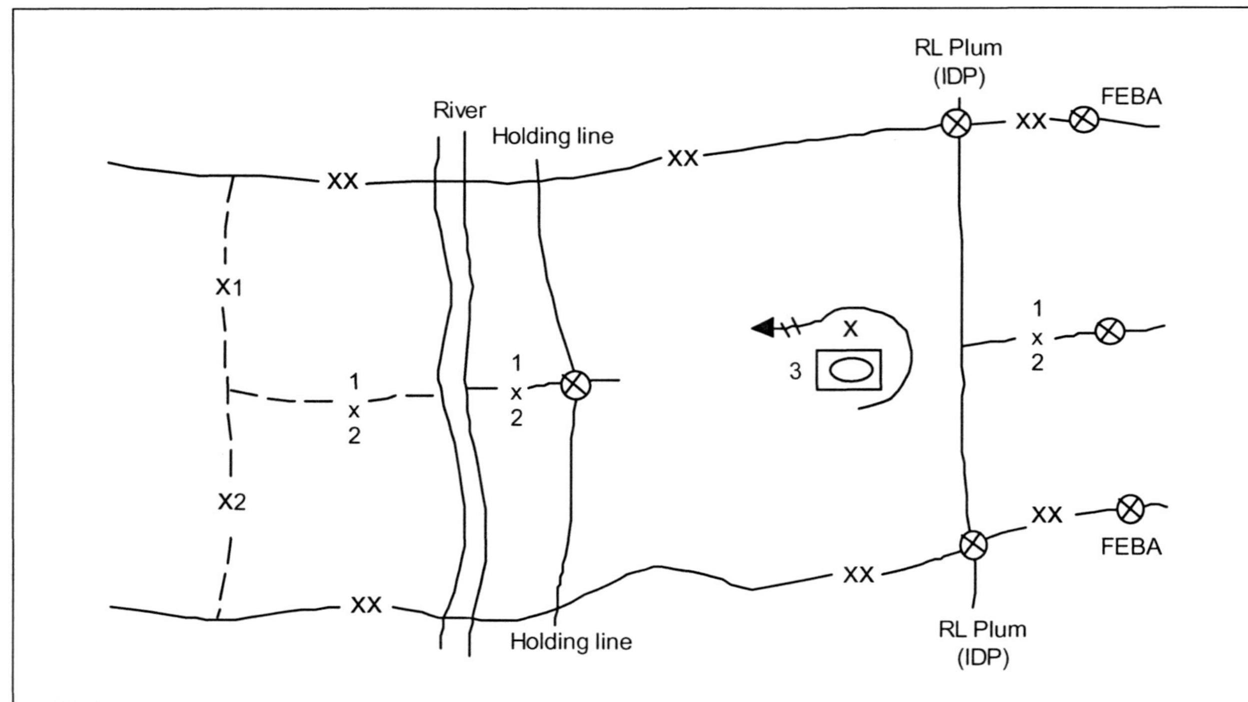

Figure 6-3. Retrograde crossing

retrograde crossing sites, which may be insufficient in number. The enemy usually knows where the logical crossing sites are and attacks them early in the operation, but it must not be allowed to capture them. Friendly forces should develop additional sites to provide flexibility against this possibility.

The commander should attempt to salvage tactical bridges and rafts for future use; however, it may be necessary to use them for the crossing and then destroy them to prevent capture. Fixed bridging must be prepared for destruction and also be protected against ground and air attacks. This requires close coordination with the delaying force to preclude cutting off friendly forces or allowing enemy seizure of sites intact.

The BMAIN, commanded by the brigade XO (CAC), is responsible for the passage of all units through the crossing area.

Traffic control up to and through the crossing area is a critical problem in crossing operations. For this reason, plans for movement must be detailed, and movement control is essential. This control is exercised by the CAC with assistance from the delaying-force commander (brigade commander). The CAC controls all movement within the crossing area to include retrograde forces.

It is the responsibility of the CAC to ensure the continuous and orderly flow of the retrograde elements across the river. His control includes both the ERPs, which ensure that all vehicles are of the proper class and size, and also all waiting areas that feed vehicles through the crossing area. To assist the CAC, MP and, if available, engineers establish and operate TCPs to manage the traffic flow. CSCs oversee the crossing means. The CAC and his staff must synchronize the crossing plan with the commander's tactical plan.

Activity within the crossing area begins with two-way crossings by CSS units evacuating nonessential supplies or restocking the delaying force. During the early stages of the retrograde, the existing crossing means may be supplemented by tactical bridging. As a minimum, additional tactical bridging assets must be planned and available.

Initially, the force crosses on fixed and floating bridges. It crosses on bridges as long as possible, since this is the most rapid means. Once the bridges become vulnerable to capture, air attack, or observed indirect fires, they may be converted to rafts or removed. Vehicles continue to cross by using rafts or by swimming. The crossings are made under the suppressive fires of the defending force's direct- and indirect-fire weapons.

The forces cross the river in an orderly flow while conserving combat power. The retrograde crossing begins as a rear-area operation for the division. Initially, it is a traffic-scheduling problem, centrally controlled by the division. The division establishes crossing areas before crossing maneuver brigades. Crossing-area operations are the same as for offensive crossings (see *Chapter 5*). Even when the division has to establish the crossing areas quickly, under adverse circumstances, it synchronizes crossing-support activities with those of the defense force that is preparing to close the routes in the crossing areas.

Crossing sites need the highest priority for AD. This is particularly critical when the enemy has air superiority or when air parity exists. The sequence for crossing AD units should account for the need to provide continuous coverage of crossing sites.

The division engineers are fully committed to the delay. As a result, engineers under the control of the CAE run the crossing sites and support initial preparation of exit-bank defenses. Engineers focus on enemy engineer breaching assets and the interdiction capabilities needed to support enemy maneuver.

Defense Phase

The defense phase stops the enemy by keeping it out of the crossing area, denying it crossing sites upstream or downstream, and destroying its attempts to cross the river. In particular, the defense phase targets potential enemy crossing assets. Whether continuing the retrograde further or defending along the river, the division establishes a strong exit-bank defense. The defending force protects the delaying force as it crosses the river after battle handover. The rearward passage of lines by the delaying force is a normal defensive operation, complicated by the river.

Initially, the defending force is small. It consists of combat and combat-support units not involved in the delay as well as augmentation from corps reserves. Because enough forces are not available to defend all points along the river, the defense depends on rapid lateral movement to concentrate at vulnerable points. In particular, it orients on and protects the crossing sites against the enemy's forward detachments and heliborne forces.

After battle handover from the delaying force, the defending force is responsible for the area between the holding line and defensive positions on the exit bank. The defending force masses fires to help its elements in contact forward of the river to withdraw, thereby complicating the retrograde crossing.

The defending force accepts battle handover from the last of the delaying force at the holding line and covers its crossing over a fixed bridge that is prepared for demolition. Friendly forces at the river prevent the enemy from crossing at the site of a demolished fixed bridge so that companies securing the crossing site can safely withdraw in turn.

WITHDRAWAL

A withdrawal differs from a delay in that it is an operation in which the unit in contact disengages from an enemy force and moves to the rear. Withdrawals are executed when the commander desires to withdraw to control future tactical operations without being forced to do so by enemy pressure. A withdrawal follows the same sequence as a delay. The only difference is that the unit may or may not be in enemy contact.

During a withdrawal, the enemy usually does not pressure withdrawing units. Also, other friendly units do not normally assist in withdrawals. Care must be taken to ensure that the enemy does not try to isolate and encircle units during river-crossing operations. If a unit has difficulty breaking with the enemy in a withdrawal, it can request help from a higher level. The assisted withdrawal will be a rearward passage of lines. Exchange of information on obstacles, indirect-fire targets, and routes in the sector must be coordinated before conducting the passage of lines. The assisting unit provides mobility support along cleared routes and corridors in its sector for the passing unit.

Engineers must complete clearing operations before the passage begins. The assisting unit also closes the lanes once passage is complete. The passing unit must plan and organize for conducting internal breaching and river-crossing operations before initiating the passage of lines. This should ensure responsive mobility operations if the enemy blocks routes during the passage.

RETIREMENT

Retirements are rearward movements away from the enemy by a force not in contact. Typically, another unit's security forces cover their movement as they conduct a tactical road march. A retirement follows the same sequence as a delay. Speed is important; therefore, engineers should focus on mobility for the retiring unit and expect operations such as route clearance and route repair.

DENIAL MEASURES

Denial measures are actions taken to hinder or deny enemy use of resources or facilities. In retrograde crossings, the commander includes bridges and crossing sites in his denial measures.

The laws of war require that denial operations, particularly against civilian resources such as existing bridges, be carefully considered and that execution authority to destroy the structure be maintained at the highest level.

A defending-force commander is responsible for preparing existing bridging and other crossing means in his sector, such as ferries, for destruction to prevent their use by the enemy. The CAE controls the engineers who prepare those targets. The timing of their destruction depends on their use in supporting the crossing. When the tactical situation dictates that crossing sites are no longer needed or the risk of capture outweighs their usefulness, the defending force must destroy them.

Use of bridges in the retrograde requires a redundant means of bridge destruction and a robust demolition guard with an engineer demolition party (see *FM 5-250*). Engineer light diving teams can be used to survey and emplace, prime, and detonate explosives on bridge supports to deny enemy access during retrograde operations. Because of the severe consequences of a premature decision to destroy a site, the division commander usually designates sites as reserve targets and issues specific orders stating under what conditions and by whose authority this destruction can be done.

Engineers destroy military bridges that they cannot recover quickly. Bridge stocks are in short supply; therefore, if existing bridges are sufficient to support the retrograde, the engineers recover military bridges early. In addition, the denial of major existing bridges can be so important that the commander may choose to destroy them early and rely on military bridges to cross the remainder of his force. The ribbon bridge is preferred for this crossing because of its recovery speed. Engineers either recover lines of communication (LOC) bridges well before the enemy arrives or destroy those left in place after the delay.

PLANNING

The division commander identifies the holding line and the units required to fight the delay and defense battles. The division engineer, in conjunction with the G3, identifies crossing sites and required crossing assets. The division staff coordinates for additional corps assets. The staff uses the planning process identified in *Chapter 4.*

The commander uses deception to conceal the extent of the operation and the actual crossing sites. Smoke, electronic warfare, and dummy sites reduce the enemy's capability to disrupt the crossing. OPSEC keeps the enemy-intelligence collectors from identifying the time and place of the crossing.

The commander may consider retaining fixed bridges in defense of the river line if he anticipates future counterattacks back across the river. He may also partially destroy bridges to ease restoration in future offensive operations, weighing this decision against the enemy's use of the bridges.

Denial operations are somewhat restrictive. Only those civilian targets with a clearly identified military value can be destroyed or removed. Coordination between the theater command and the host-nation government is important in the policy-development process.

CHAPTER 7
Crossing Sites

GENERAL

The following paragraphs supplement the general descriptions of acceptable crossing sites in *Chapter 2.* Selection of crossing sites is primarily based on the—

- Existing situation and anticipated scheme of maneuver.

- Physical characteristics of the available sites, road networks, and surrounding terrain.

- Availability and capabilities of crossing means.

- Availability of engineer support.

Conflicts between tactical and technical requirements frequently occur. Commanders evaluate the factors bearing on the problem to determine the best overall solution.

CROSSING-SITE SELECTION

Each crossing means, except air assault, requires a type of crossing site. They can be identified as fording, assault-boat, swimming, rafting, or bridging sites. Assault battalions use either a fording or an assault-boat site (or sometimes a swimming site) as an assault site.

Both the desired scheme of maneuver and available crossing means influence crossing-site selection. The division assigns a crossing area to each lead brigade. The brigade chooses which crossing sites to use within its area. When a particular site is important to the division's tactical concept, such as for the movement of breakout forces, the division either coordinates with the affected brigade to open that bridge site or moves a bridge to that site once the brigade hands over the crossing area to the division.

Brigade commanders select final crossing sites based on tactical intelligence and their desired schemes of maneuver. Each site's physical characteristics, required engineer support, and available crossing means influence the decision, but tactical requirements are the most important.

The goal when selecting assault sites is to pick those that allow the lead battalions to cross unopposed and seize far-shore objectives rapidly. If unsuccessful at finding undefended crossing sites, the lead battalions cross under enemy fire while overwatch units provide direct and indirect suppressive fires. Assault sites may or may not coincide with rafting or bridging sites.

When selecting swimming sites, the goal is to pick those that permit fighting vehicles to rapidly enter, swim across, and exit the water with minimum assistance.

The goal when selecting rafting and bridging sites is to pick those that support the greatest volume of vehicle traffic consistent with the scheme of maneuver. Rafting and bridging sites are usually on or near major roads to minimize route preparation and maintenance. When the sites are located close together, the bridging site must be upstream of the rafting site. This will avoid potential damage that may be caused by disabled rafts drifting into the bridge.

Regardless of the crossing means, each site needs engineer reconnaissance swimmers or an engineer light diving team to cross early, reduce obstacles, and develop exit points on the far bank. Riverbanks at otherwise suitable crossing sites often need work for access to the river. Most natural soil becomes unstable under heavy traffic. This condition worsens as fording, swimming, and rafting activities carry water onto it. The required engineer effort varies with soil type, crossing means, and vehicle density. An engineer vehicle that is capable of maintaining the exit bank should be one of the first vehicles across.

Natural conditions vary widely. Banks may require little preparation, or they may be so restrictive that they limit feasible sites. Desirable site characteristics include—

- Minimum exposure to enemy direct-fire weapons.

- Covered and concealed access to the river's edge.

- Firm and gently sloping banks that allow rapid entry and exit at multiple points.

Initial and subsequent entry points can vary. Available locations seldom have all the desired tactical and technical characteristics. The best routes through the crossing area normally cross the river at the technically best crossing sites. The best technical sites may not be the best tactical sites because they are well known and are heavily defended by the enemy. Forces initially crossing at less desirable locations are most likely to avoid detection and gain surprise. Moving laterally along the exit bank, forces attack the flank or rear of enemy units to seize the better crossing locations. Use of these sites allows rapid buildup of combat power.

PLANNING

Planners need information of potential crossing sites to evaluate their compatibility with proposed crossing plans. Generally, planners need to know—

- Friendly and enemy capabilities and probable COAs.

- Site capacity for the crossing of troops, equipment, and supplies using various crossing means.

- Engineer support that is required to develop, improve, and maintain each site.

More specifically, planners need to know the—

- Condition of the bottom, banks, and water of the river.

- Impact of forecasted or past seasonal weather conditions.

- Location of defensible terrain, covered and concealed areas, and natural or enemy-emplaced obstacles on both sides of the river.

- Amount of time and effort that is required to develop sites, assemble rafts, and construct bridges.

- Entry/exit routes and off-road trafficability.

- Road networks.

- Capabilities of friendly forces to deny observation, suppress fires, and provide site protection.

REQUIREMENTS

Discussed in the following paragraphs are some of the requirements necessary for crossing sites.

ENTRY/EXIT ROUTES OR PATHS

A desired feature of all sites is readily accessible entry/exit routes or paths on either bank. The approaches to the banks are checked for their ability to support the requirements (width, slope, and trafficability) of the wheeled and tracked vehicles of the crossing element. Covered and concealed approaches enhance surprise and survivability; however, multiple routes, free from obstruction, will increase crossing speed and flexibility. Exit-bank conditions often take precedence over entry-bank conditions until equipment and troops can be crossed to develop and improve the site. See *Table 7-1* for depth requirements.

ROUTES AND APPROACHES

Depending on the crossing operation that is used, the following considerations must be given to routes and approaches:

- Fording. Dismounted forces may use approaches with steep slopes and heavy vegetation, while vehicle fording requires paths or roads to approach fording sites.

- Assaulting or swimming. Assault-boat crossings may use more rugged approaches than amphibious vehicles.

- Rafting. Multiple approach routes to rafting sites permit the relocation of rafting upstream or downstream.

- Bridging. Bridging sites require developed road networks to sustain the crossing capacity.

Depending on the vehicle that is used, the following considerations must be given to routes and approaches:

- Wheeled vehicles. In general, wheeled vehicles require 3.5-meter path widths and 3.5 meters of overhead clearance. Dry, hard slopes of 33 percent can be negotiated; however, slopes less than 25 percent are desired.

- Tracked vehicles. Tracked vehicles require up to 4-meter path widths and 3.5 meters of overhead clearance. Tanks can climb 60 percent (31-degree) slopes on dry, hard surfaces; however, slopes less than 50 percent are desired.

WAITING AREAS

Numerous waiting areas are required for equipment and troops preparing and protecting sites and for troops and vehicles preparing and/or waiting to cross. These areas should be dispersed, provide cover and concealment, and be accessible to road networks near the sites.

Table 7-1. Depth requirements

Operation	Item	Draft in Meters	Remarks
Fording	Personnel	0.10	
	Wheeled vehicle	0.75	
	Tracked vehicle	1.20	
Assaulting/ swimming	M113 (APC)	1.50	The water should be deep enough for the vehicles being used.
	M548	2.00	
	M2	1.07	
Rafting/bridging	Power boat (27 feet)	1.00	The water should be deep close to the bank to preclude grounding of the raft or bridge.
	LTR	0.60*	
	M4T6	0.75*	
	Ribbon	0.60*	
NOTE: *Power boat draft may govern depth.			

RIVER CONDITIONS

In general, currents less than 1.5 meters per second (MPS) are desired. Narrow segments of the river decrease equipment requirements, crossing time, and exposure time. However, resulting increased current velocities may offset any advantage. As the current's velocity increases, it decreases the ribbon bridge's ability to handle heavy military load class (MLC) vehicles. More boats will be required to keep the bridge in place and allow for heavy MLC vehicles to cross.

BANKS

Ford banks may be steep and rugged for dismounted troops; however, vehicles require slopes less than 33 percent and firm soil conditions. Assault or swim banks may be steep when using assault boats for dismounted troops. Amphibious vehicles may be able to enter over low, 1-meter vertical banks, but they require sloped exits. Vertical banks of about 1 meter may be accommodated by bridge or raft ramps (see *Table 7-2*). Vertical bank heights for bridges using the equipment listed in the table do not change for ribbon bridges. For M4T6 and Class 60 bridges, the height of the bridge deck can be adjusted to accept a difference in bank heights; however, the limiting factor may become the longitudinal slope of the bridge.

BOTTOMS

Ford bottoms must be free from obstacles, firm, and uniform. Riverbeds may be improved with rock fill or grading equipment. Guide stakes make the crossing of the river easier for boat drivers. Assault- or swim-site bottoms must be free from obstructions that interfere with boats or the tracks of amphibious vehicles. Rafting sites must be free from obstructions that could interfere with boat operations. Bridges emplaced for lengthy periods (4 hours or more) or in strong currents require suitable riverbeds for anchorage. Engineer light diving teams from the corps Army may be used to—

- Conduct river-bottom reconnaissance.

- Emplace shore and midstream anchorage for debris and antimine and antidiver nets to ensure the success of the operation.

ENEMY SITUATION

Sites masked from enemy observation enhance surprise and survivability by degrading the enemy's ability to see. Using existing sites reduces preparation time but requires caution in that the enemy may have emplaced obstacles and registered artillery on the site.

Table 7-2. Bank requirements

Equipment	Ramp Articulation (raft)	
	Up	Down
M4T6	Not adjustable	
Class 60	Not adjustable	
Ribbon	1.7 meter*	

NOTE: *An approach ramp 2.1 meters (7 feet) long provides extra roadway for loading and off-loading vehicles.

SITE ANALYSIS

A ground reconnaissance refines and confirms information gathered from other sources. *Training Circular (TC) 5-210* contains details for conducting and reporting site reconnaissance. From this and other detailed reports, planners may develop charts or overlays to compare alternate sites. Unit SOPs may prescribe specific comparative methods. See *Figure 7-1* for an example.

Bank Preparation Time

- Describe the height, slope, and stability of the bank.
- List the amount of time and effort required to overcome significant natural and enemy-emplaced obstacles.
- Include day, night, or other reduced-visibility constraints.

River Conditions

- Specify the width, depth, velocity, and bottom conditions of the river, as appropriate.
- Include variations or unique factors (such as sandbars, turbulence, or water depth at the bank).

Vegetation

- List areas suitable for work sites and AAs and available cover and concealment.

Full Crossing Rate

- Describe foot-, wheeled-, and tracked-movement capabilities on roads, trails, and cross-country.
- Describe the maximum crossing rate for fording, swimming, or rafting.
- Include the overall assessment of the crossing site's potential.

Figure 7-1. Crossing-site requirements

FIELD CALCULATIONS

Some common relationships and field-expedient calculations that are useful during a ground reconnaissance include—

- Measuring the current's velocity.
- Determining slopes and degrees.
- Measuring the river's width.
- Calculating downstream drift.

MEASURING THE CURRENT'S VELOCITY

Correlating the desired maximum current velocity of 1.5 MPS with a familiar comparative unit of measure may help in estimating the current's velocity. The quick-time march rate of 120 steps per minute, with a 30-inch (or 76-centimeter) step, equates to 1.5 MPS. Other approximate correlations of 1.5 MPS include—

- 5 feet per second (fps)
- 3.5 miles per hour (mph)
- 5.5 kilometers per hour (kph)

Determining the current's velocity is critical to effective and safe crossing operations. When it is high, more boats are required to stabilize the bridge, particularly when anchorage is not used. A reasonable estimation involves measuring a distance along the riverbank and noting the time a floating object takes to travel the same distance.

Dividing the distance by the time provides the current's velocity (see *Figure 7-2*).

DETERMINING SLOPES AND DEGREES

The slope of terrain is significant (for example, slopes of 7 percent or more slow movement and may require vehicles to operate in a lower gear). Slope, usually expressed as a percentage, is the amount of change in elevation (rise or fall) over a ground (horizontal) distance (see *Figure 7-3*).

The means to determine the percent of the slope include—

- Clinometers. These instruments measure the percent of the slope and are organic to most engineer units down to the platoon level.

- Maps. In this method, first measure the horizontal distance along the desired path, then determine the difference in elevation between the starting and ending points of the path. The next step is to ensure that both figures are the same

unit of measure (such as meters or feet). The final step is to divide the elevation (rise) by the distance (run) and multiply the result by 100 to get the percent of the slope (see *Figure 7-4*).

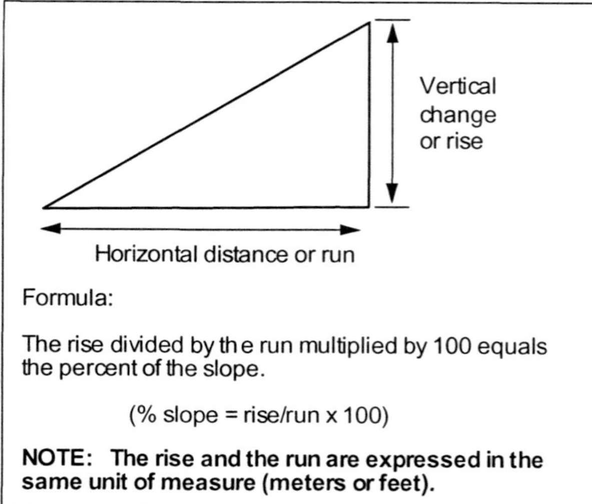

Figure 7–3. Slope calculation

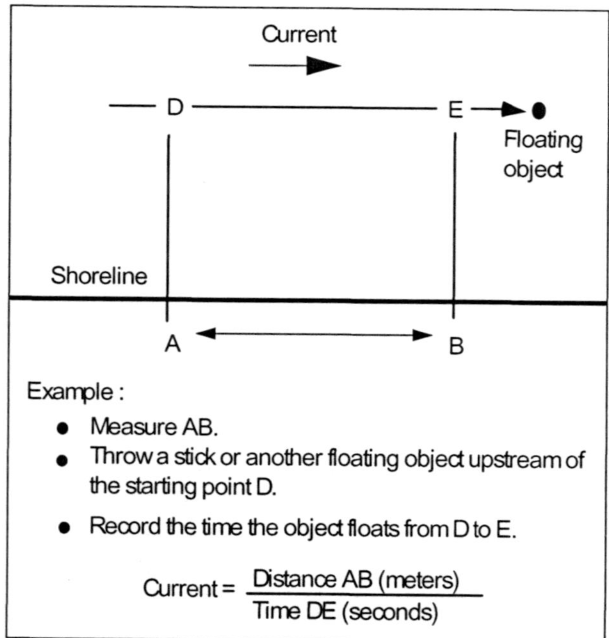

Figure 7–2. Measuring the current's velocity

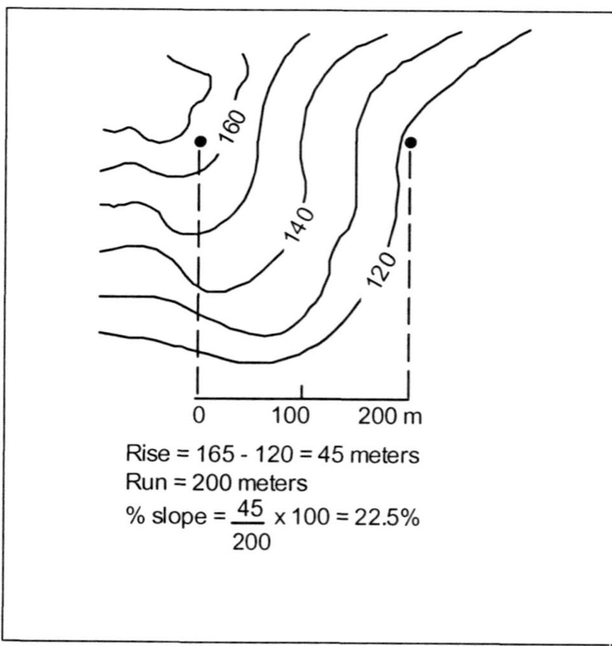

Figure 7–4. Terrain slope

- Line of sight and pace. This method uses the eye-level height above ground (usually from 1.5 to 1.75 meters) and the length of standard pace (usually 0.75 meters). While standing at the bottom of the slope, the individual picks a spot on the slope while keeping his eyes level. He paces the distance and repeats the procedure at each spot. Adding the vertical and horizontal distances separately provides the total rise and run.

Slope may also be expressed in degrees; however, this provides angular measurements. The method is not commonly used because the relationships are more complex than desired for field use. *Figure 7-5* lists some relationships of the percent of the slope to the degree of the slope.

MEASURING THE RIVER'S WIDTH

A field-expedient means of measuring the river's width is with a compass. While standing at the waterline, fix your sight on a point on the opposite side and note the magnetic azimuth. Move upstream or downstream until the azimuth reading to the fixed point on the opposite bank is 45 degrees different than the original reading. The distance from the original point to the final point of observation is equal to the river's width (see *Figure 7-6*).

CALCULATING DOWNSTREAM DRIFT

Current causes all surface craft to drift downstream. Each vehicle has a different formula for calculating downstream drift. Amphibious vehicles and assault boats drift more

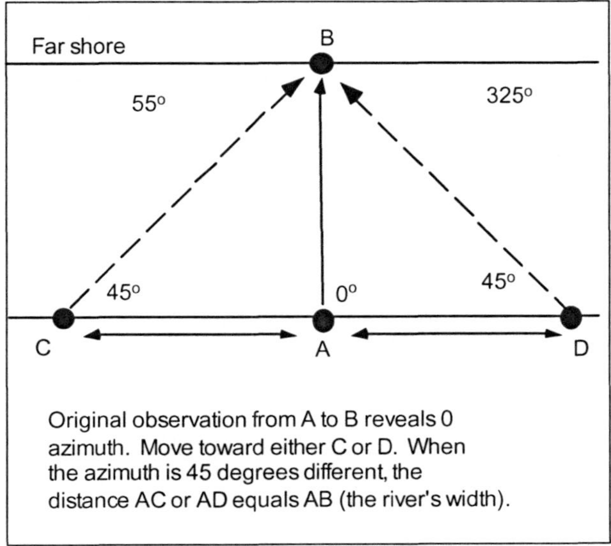

Figure 7–6. River's width

than powered boats and rafts; the latter has a greater capability to negate the effect of the current's velocity by applying more power.

Amphibious vehicles and nonpowered assault boats are generally limited to current velocities of 12.5 to 2 MPS and 1 MPS respectively (see *Figure 7-7*).

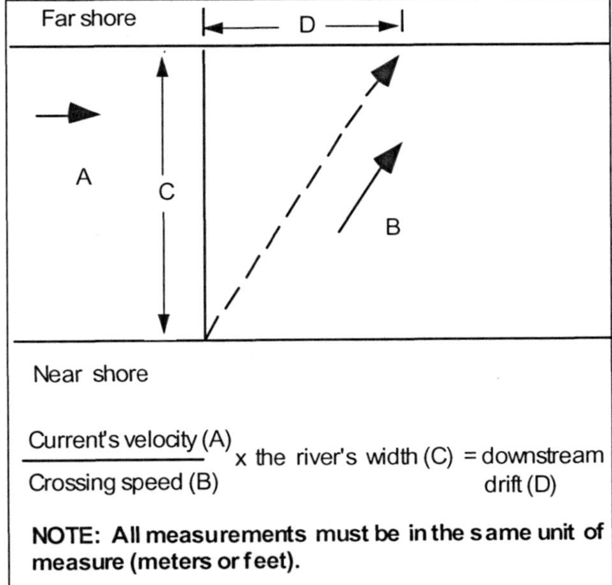

$$\frac{\text{Current's velocity (A)}}{\text{Crossing speed (B)}} \times \text{the river's width (C)} = \text{downstream drift (D)}$$

NOTE: All measurements must be in the same unit of measure (meters or feet).

Figure 7–7. Amphibious drift

Slope	Degrees
100%	45
60%	31
40%	22
20%	11

Figure 7–5. Relationship of slopes to degrees

When crossing with amphibious vehicles and pneumatic boats, compensate for the effect of the current (see examples below).

Example 1

Entry is usually made upstream of the desired exit point. The vehicle or boat is aligned, or aimed, straight across the river, creating a head-on orientation that is perpendicular to the exit bank. However, the current produces a sideslip, downstream forward movement (see *Figure 7-8)*. This technique requires operator training in continual adjustment to reach the objective point on the exit bank. This technique results in a uniform crossing rate in the least amount of time and is usually the desired technique.

Example 2

If the operator continues to aim the vehicle at the desired exit point, the orientation of the craft at the exit point will approximate an upstream heading. The craft's path is an arc in proportion to the current's velocity (see *Figure 7-9)*.

Example 3

To exit at a point directly across from the entry point requires an upstream heading to compensate for the current's velocity (see *Figure 7-10)*.

In all three examples, the craft's speed relative to the current's velocity is constant, assuming the engine revolutions per minute (rpm) or paddling rate remains constant.

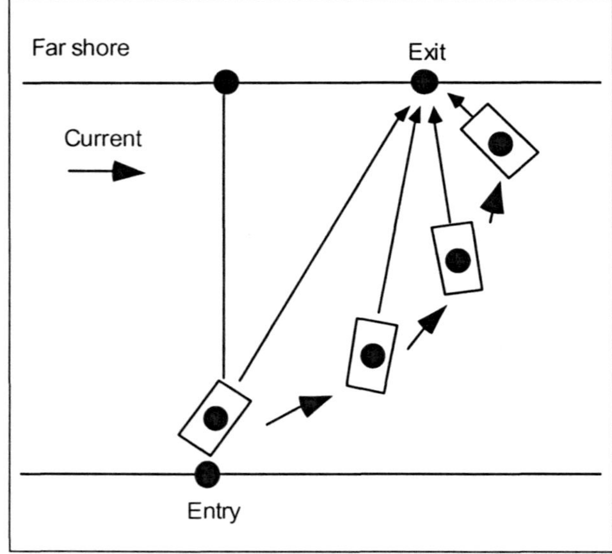

Figure 7–9. Constant aim point

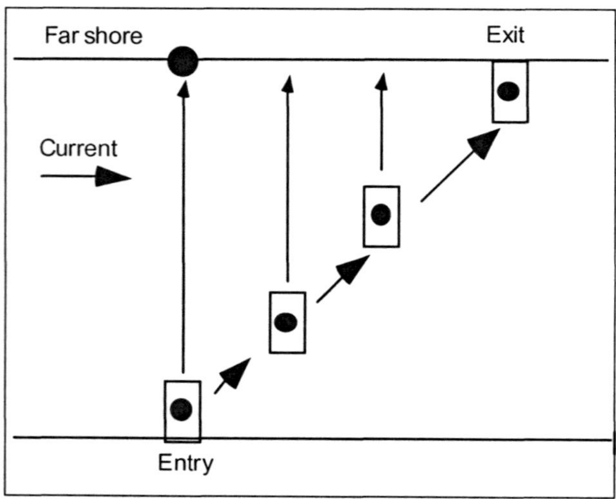

Figure 7–8. Downstream sideslip

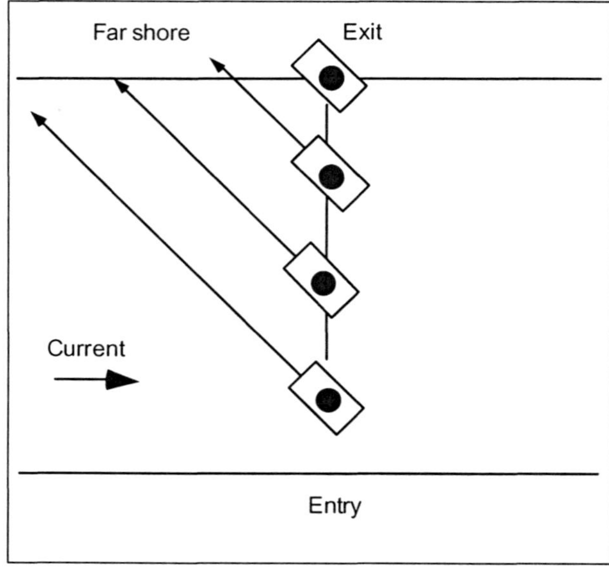

Figure 7–10. Constant heading

Terrain conditions may restrict the location of entry and/or exit locations. Enemy situations may require alternate techniques. For example, when aiming at the downstream exit point, the craft moves at a greater speed relative to the banks after entry than it does as it nears the exit due to the current's velocity. Use of this technique may be favored when the enemy has a better observation of the entry bank rather than the exit bank. Watercraft moving fast and at a changing rate are more difficult to engage effectively.

RAFTS

The assault force seizes a far-shore objective and then clears in zone to secure the crossing sites from direct fire. Quick reinforcement with armored fighting vehicles is critical when the initial assault is dismounted.

Given the vital need to rapidly build combat power on the far shore, the lead brigades should swim the fighting vehicles of the follow-on battalions whenever practical to save the rafts for the tanks. Rafts are usually the initial means for crossing non-swimming vehicles, particularly tanks, on wide, unfordable rivers. It may be possible to bridge immediately after the assault across the river; however, rafting is normally first because rafts—

- Are less vulnerable to enemy air and indirect fire due to their size and maneuverability.

- Are quicker to assemble.

- Offer more flexibility in operation, particularly in site selection and subsequent movement between sites.

- Can use existing road networks and banks where access and exit routes are not aligned opposite of each other.

Raft assembly begins on order, not according to a preplanned schedule, even though the crossing plan has an estimated start time. Unless the division commander directs otherwise, the brigade commander, advised by his CAE, decides when to begin rafting. This is always after eliminating enemy direct fire on the site and usually after neutralizing observed indirect fire. Massed enemy indirect fire can cover an entire rafting site. The force neutralizes observed indirect fire by suppressing it and obscuring the crossing sites.

SITE PREPARATION

The brigade commander also decides when bank preparation can begin, or he may delegate this decision to his CAE. The decision is based on the estimated time required to secure the area. By initially spending extra time and effort preparing a bank, maintenance problems can be avoided during rafting operations.

The key to rapid and effective bank preparation is good engineer reconnaissance, which permits engineers to arrive at the site early, organized and equipped to perform specific tasks to improve the approach. The same is true on the exit bank. Poor bank conditions require early priority for raft movement of engineer equipment across the river. Time spent preparing the exit banks before passing heavy traffic greatly reduces maintenance of the crossing site and speeds force buildup later. Two entry and exit points per centerline make it possible to alternately use one while maintaining the other.

RAFTING SITES

Each lead brigade should have at least two rafting sites, each of which has one to three raft centerlines. The terrain, routes, and tactical plan determine their location. However, they should not be closer together than 300 meters to avoid congestion and the risk that enemy artillery concentrations will impact on more than one site during a barrage. Engineers prepare alternate sites as soon as possible to permit the relocation of rafts in case of enemy action or bank deterioration. Each rafting site also contains the control and operation structure that is necessary to conduct rafting operations (see *Figure 7-11*).

Engineer platoon leaders are in charge of rafting sites. Each site has one to three active centerlines spaced 100 to 300 meters apart. With the 100-meter minimum distance between centerlines, collisions between rafts on adjacent centerlines are avoided and the effects of artillery are reduced. However, spacing centerlines farther than 300 meters apart stretches the ability of one's unit to control both land approaches and water operations. Each crossing site has at least one alternate centerline. The CSC switches to the alternate centerline when necessary due to enemy fire or bank maintenance.

The centerline has an embarkation point on the near bank, a debarkation point on the far bank, and rafts operating between these two points. The number of rafts on a centerline depends on the river's width and the unit's control (see *Appendix C*). Maintaining bridge-unit integrity on centerlines and crossing sites is critical. It

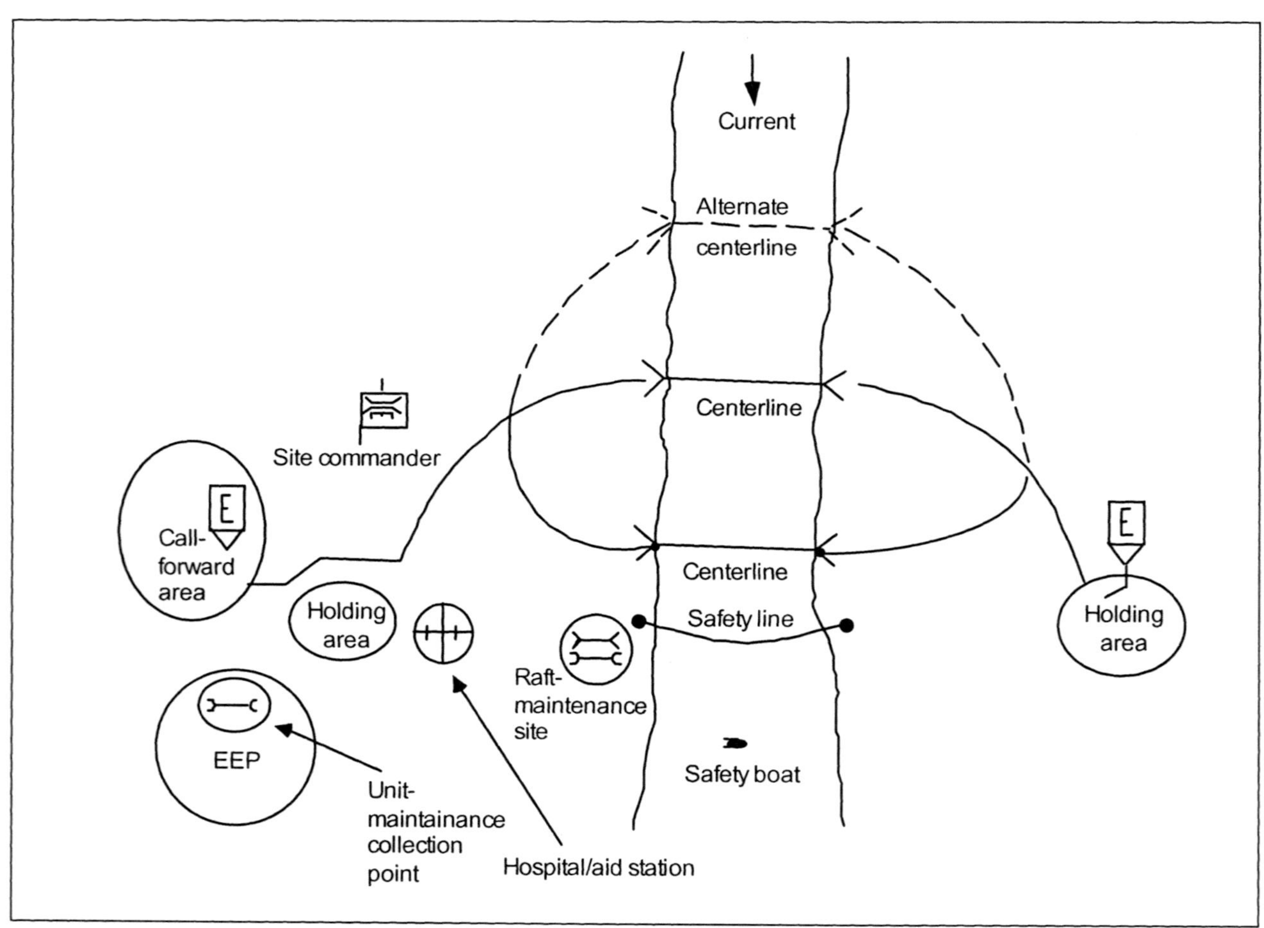

Figure 7–11. Rafting site

simplifies the maintenance and operation of rafts and significantly improves control on the water, as all raft commanders and boat operators have trained together. On any centerline, rafts must be the same type and configuration.

Centerlines are marked to guide vehicles approaching and leaving the water and to guide rafts to the correct landing points. Marker stakes or panels are used during daylight, and dim lights (covered flashlights or chemical lights) are used at night (see *Figure 7-12*). The raft commander designates the location of the markers depending on the terrain and the current's velocity. Markers include the following:

- Raft-guide markers, at a 45-degree angle upstream, are used to guide the raft to

the embarkation or debarkation point. The two markers are 91 centimeters apart, and the marker farthest from the river is 60 centimeters higher than the other. The raft has the correct approach to the bank when the markers appear to be in a straight line, with one above the other.

- Raft-landing markers depict the left and right limits of the embarkation or debarkation point.

- Vehicle-guide markers are used to align raft loads with the raft and are visible to both the raft and the vehicles.

Each rafting site contains at least one safety boat, normally a bridge-erection boat (BEB), for troop and equipment recovery. The bridge

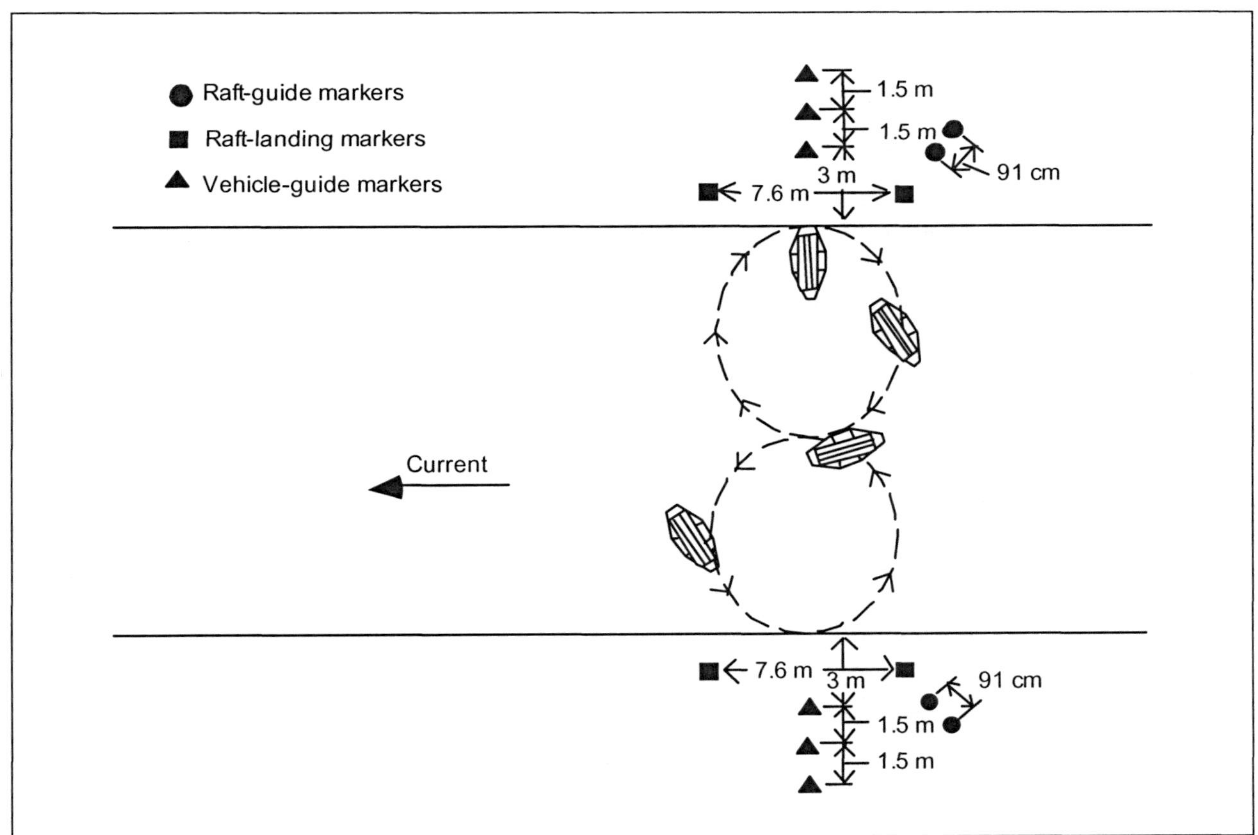

Figure 7–12. Centerline marking and operation

company provides a crew for the safety boat, including a boat operator, a boat commander, a medic, and a lifeguard (two, if possible). The lifeguard-qualified swimmer does not wear boots or load-bearing equipment (LBE). The safety boat also has a float with an attached line for rescuing troops in the water, a boat hook, rocket-propelled lifelines (if available), and night-vision goggles (for the boat commander, primarily). It has a radio on the bridge company net. The safety boat maintains its station 50 meters downstream of the rafting site.

As soon as possible, a safety line should be run across the river 100 meters downstream from the last centerline. This line is fastened to the banks and kept afloat by life jackets attached to the line every 30 meters. This rope acts as a catch rope for troops who may fall overboard during rafting operations, especially during limited visibility. The safety line does not replace the need for a safety boat but helps in case several soldiers fall into the river.

Each crossing site requires an EEP located where the equipment will be accessible to the crossing site. Traffic between the EEP and the riverbank should use a separate route to avoid congestion with the crossing.

Each rafting site requires a place along the friendly shore, downstream of the centerlines, for immediate raft repairs. The maintenance area requires an access point to the river for the removal and launching of bays and boats. Additional equipment desired at the maintenance area includes—

- A bridge boat to move damaged bays and serve as a spare boat.

- A crane to remove nonrepairable equipment from the water.

- A bridge truck to transport damaged equipment to the EEP.

- A fuel heavy-expanded mobility tactical truck (HEMTT) to refuel boats.

- One or more interior and exterior bays to use as replacement parts.

The maintenance area is continuously manned with—

- Two mechanics with tool boxes.

- Two fuel handlers.

- The operators of the various pieces of equipment.

- A site supervisor.

RAFTING OPERATIONS

Units begin their preparations for rafting operations at a staging area. There, they receive briefings, conduct inspections, and rehearse the rafting operation. Personnel will be issued life jackets and given instructions on what to do upon loading onto the raft.

When ordered to begin rafting, the site commander directs personnel at the ERP in the call-forward area to begin sending raft loads forward. Units proceed from a staging area to the call-forward area where engineers at the ERP organize them into raft loads and send them down to the river. Any points along the route that may cause confusion, such as intersections, are either manned with a guide or are marked to ensure that the vehicles do not get lost. Once a raft load nears the river, the platoon leader directs it to the appropriate centerline. The platoon leader controls the flow of traffic to the centerlines to ensure that there is a smooth flow of traffic and that centerlines are neither congested nor underused. He establishes the timing required so that raft loads leave the call-forward area and match up with a returning empty raft.

When a raft load reaches the riverbank, it is met by an engineer centerline guide. He stops the raft load 3 meters from the edge of the water and holds it there for the raft commander. The raft commander guides the vehicles of the raft load onto the raft. The raft crew chocks the vehicles and ensures that all passengers are wearing life jackets. The passengers do not dismount from their vehicles. All hatches are opened to allow quick exit of the vehicle in case of an emergency. Upon reaching the debarkation point, the raft commander guides the vehicles off the raft. After the raft load debarks, the raft commander checks with the centerline guide for any return vehicles and returns to the embarkation point.

Once on the far shore, the centerline guide directs the raft load to the far-shore holding area where it re-forms. The passengers remove their life jackets, which are collected and returned by an engineer team to the staging area for future loads.

MAINTENANCE AND REFUELING

During rafting operations, rafts require stops for refueling, preventive maintenance, and minor repairs. The efficiency of the crossing depends on all rafts having enough fuel and minimal lost time for refueling and normal maintenance. This efficiency requires the bridge company to intensely manage raft maintenance and to operate the maintenance area much like a pit crew in an automobile race. When directed, a raft pulls off the centerline and moves to the crossing-site maintenance area.

With the raft secured, the crew begins refueling and maintenance operations. Mechanics assess and repair any minor damages to the raft and the boats. Fuel handlers run fuel lines from the fuel to both bridge boats and fuel them simultaneously. If no major deficiencies are identified, the entire process requires 20 minutes. If major deficiencies are identified on the boat, it is removed from the raft and replaced with an awaiting spare. The boat will then be removed from the water and sent back to the EEP for repair. When refueling and maintenance operations are finished, the raft returns to its centerline and another raft is directed in for maintenance and refueling.

Since the maintenance and refueling operation is continuous and requires removing a raft from the operation for up to 30 minutes, it is important to account for this reduction in capabilities when planning the operation. Generally, it is unnecessary to refuel for the first two hours after rafting begins. Once raft maintenance and refueling begin, one of the six rafts in each bridge company is unavailable for carrying vehicles across the river at all times.

When a raft becomes damaged and needs immediate repair, the raft commander moves it to the maintenance area. If a raft loses a boat and cannot make it to the maintenance area without assistance, the raft commander contacts the maintenance supervisor, who sends the maintenance boat out to assist. If a raft is still carrying a load, the raft commander decides on which bank he will disembark the load. Once in the maintenance area, mechanics determine the extent of the damage. If the damage requires significant repair, the raft will be removed and replaced with a spare. Lengthy equipment repairs are done at the EEP.

BRIDGES

Bridges replace or supplement rafts once enemy-observed indirect fire is eliminated. Each lead brigade should convert at least one rafting site to a bridge site as soon as possible, while keeping other rafting sites in operation until a second bridge is in place. Bridges have greater traffic-flow rates than rafts. The ribbon bridge is the preferred initial bridge, since it is faster to assemble and easier to move than other types. Once assembled, all float bridges have a crossing rate of 200 vehicles per hour, with a vehicle speed of 15 mph. As with rafts, bridge assembly begins on order, not according to a preplanned schedule. Since vehicles cross rivers much faster on bridges than on rafts, early bridge assembly is desirable but must be weighed against the risk that the enemy can still bring indirect fires down on an immobile bridge. The brigade commander decides when to begin, with advice from his engineer. He may delegate this decision to his CAE.

Bridges need protection. AD, counterfires, and ground-security elements are necessary to defeat enemy attacks. Booms on the river protect bridges from damage caused by waterborne munitions and debris. Antidiver nets are placed upstream and downstream to protect bridges from swimmers or underwater demolition teams. Engineer light diving teams may be employed to reduce debris along the debris-collection side of the upstream boom using portable hydraulic chain saws.

Bridges are vulnerable to enemy long-range artillery fire and air attack even after the assault force clears enemy forces from the exit bank. For this reason, ribbon bridges are used for a limited period of time, normally two to four hours, before engineer bridge units break them apart and move them to other sites. When the division uses this pulse-bridging tactic, its units wait to cross in staging areas and surge across when bridges are in place.

Enemy air superiority over the river may prohibit bridge assembly. A sustained enemy air attack forces engineers to break established float bridges into rafts. This minimizes the destruction of scarce bridging assets yet enables the crossing to continue, though at a slower pace. Engineers prepare

alternate sites and position spare equipment nearby in case of enemy action.

As the danger from enemy action lessens, engineers use the more slowly assembled LOC and M4T6 bridges to augment and then replace the tactical bridges (ribbon bridge or AVLB). They do this as soon as possible to move ribbon bridges forward to other crossing operations.

Enemy bridges captured by the lead brigades are a bonus and speed the crossing. Engineers with the lead brigades neutralize explosive devices and reinforce weak or damaged bridge structures whether the damage is above or below the waterline. Commanders rarely base the success of an operation solely on the seizure of intact bridges.

SITE ORGANIZATION

A bridging operation requires a continuous traffic flow to the river. Units must be quickly briefed and moved to the crossing site. To accomplish this, units receive briefings in the staging areas from traffic-control personnel. There is no intermediate call-forward area. To control crossing vehicles, the engineers from the bridge unit set up an ERP at the bridge's access points on each side of the river. These engineers guide vehicles onto and across the bridge, ensure the proper speed and spacing of vehicles on the bridge, and prevent vehicles too heavy for the bridge from trying to cross.

A recovery team is stationed on the far shore to remove any damaged vehicles from the bridge. The recovery team consists of a medium or heavy recovery vehicle and crew, with sufficient winch cable to reach across the bridge. A typical site setup is shown in *Figure 7-13.* The bridge site must have several possible centerlines with adequate road networks from the unit's staging areas. Individual centerlines are spaced greater

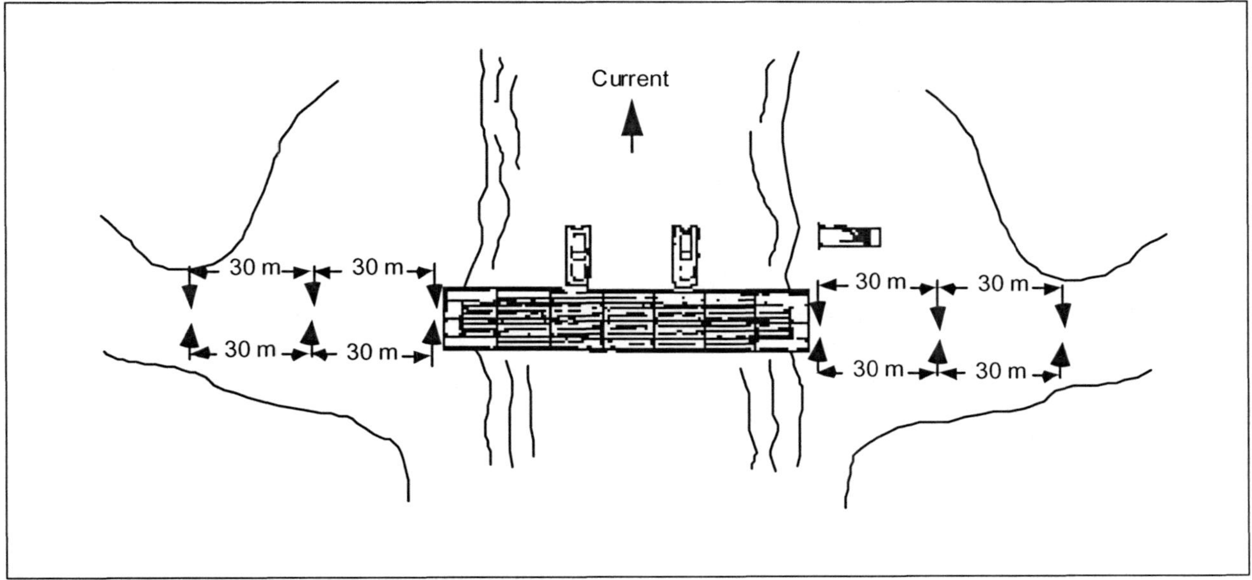

Figure 7–13. Bridge-crossing site

than 300 meters apart to reduce the effect of enemy artillery and air attacks.

Any method can be used to mark the route to the bridge as long as markers are visible to the operators of the vehicles and are masked to observation from above. As the vehicle approaches the bridge's edge, markers are spaced 30 meters apart to assist operators in visualizing the required vehicle interval on the bridge.

NIGHT OPERATIONS

At night and during limited visibility, bridge company soldiers guide the vehicles across the bridge to prevent them from driving over its side and to help them to maintain the correct crossing speed.

WARNING

Soldiers must not ground guide vehicles by walking in front of them. Instead, soldiers should be placed as guides in stationary positions at reasonable intervals along the bridge.

Guides carry flashlights or chemical lights to guide the vehicles. The first guide momentarily stops the vehicle at the ramp and guides it onto the bridge. He then shields the lights with his body and steps out of the roadway. The second guide, spaced about 30 meters farther along the bridge, unshields his lights and directs the vehicle to his location. When the vehicle operator sees a guide shield his lights, he looks further down the bridge to pick up the next lights and is guided to them. When the vehicle reaches the guide, he shields his lights and steps aside so that the next guide can pick up the direction of the vehicle. This procedure is continued across the bridge.

ACTIONS UNDER FIRE

If the unit comes under fire while on the bridge, those vehicles that are on it continue moving to the other side and leave the area. Vehicles that are not yet on the bridge stop and go into a herringbone formation or take up concealed positions. Once all vehicles have cleared the bridge, the bridge crew will break it into rafts and disperse them to reduce their vulnerability to incoming fire.

VEHICLE RECOVERY

If a vehicle breaks down on the bridge, the bridge crew will immediately attach a winch cable from the far side and drag the vehicle off the bridge. The recovery vehicle will not move onto the bridge and tow the disabled vehicle off since the critical requirement is to clear the bridge and maintain traffic flow; loss of the vehicle is far less important.

OTHER GAP-CROSSING EQUIPMENT

The AVLB allows the maneuver force to conduct a hasty short-gap crossing without requiring bridging assets from the corps. The AVLB allows vehicles up to 70 tons to cross over a 15-meter gap. During the launch, the AVLB presents a high profile that will reveal the breaching location. The bridge can be launched on the near shore and then recovered on the far shore and used again. Additionally, the AVLB can be overlaid on an understrength bridge allowing heavy tracked vehicles to cross. In such instances, the AVLB must be cribbed to avoid damage to the bridge. The AVLB can be placed on a stream bottom to assist vehicles fording over soft or rocky material, but caution must be used not to cause severe damage or bottom suction, rendering the bridge unrecoverable.

Overlapping several bridges and interlocking them can allow vehicles to ford rivers of greater width, but this works best when the only limiting factor is the depth of the river. In the near future, the AVLB will be replaced by the Wolverine, which can provide a 24-meter gap-crossing capability for MLC 70 vehicles. It will be transported, launched, and retrieved by an Abrams chassis with the turret removed.

The primary role of the MGB, a hand-erectable, heavy-duty bridge, is for tactical bridging in the forward main battle area. The key advantage of the MGB over other fixed bridges is its speed and ease of erection and little, if any, site preparation. During assembly, soldiers will be vulnerable to small-arms fire, indirect artillery fire, direct weapons fire, and a nuclear, biological, chemical (NBC) environment. As the situation permits and the enemy threat is reduced, the MGB would replace the AVLB to allow greater traffic across the gap. Eventually, the MGB would be removed to be relocated forward and replaced by other standard or nonstandard bridging.

As the danger from enemy action lessens, engineers use the more slowly assembled LOC bridges to augment and then replace the tactical bridges. The primary use of the Bailey bridge is a temporary LOC bridge. It can be used in forward areas to replace assault bridging and the MGB. This bridge system can also be assembled as a railway bridge, thus providing a relatively rapid repair capability. In some cases, the Bailey bridge is the only tactical bridge suitable for long spans and heavy loads because it can be assembled in multiple heights and widths. Currently, there are large operational stocks of the Bailey bridge in Europe, but there are no plans for additional procurement. Many allied nations have fielded the Compact 200 Panel Bridge that is similar to the Bailey but uses advanced alloys which give it greater strength.

Enemy bridges captured by the lead brigades are a bonus and speed the crossing. Engineers with the lead brigades neutralize explosive devices and reinforce weak or damaged bridge structures. Commanders rarely base the success of an operation solely on the seizure of intact bridges and should not incorporate it in a site-crossing plan. Targeting of such bridges for direct or indirect fire is relatively simple as both sides know the locations. In many cases, the load classification of civilian bridges is not enough for heavy vehicles such as tanks. If used in defensive or retrograde crossings, civilian bridges should supplement other crossing techniques.

CHAPTER 8
Assault Crossing

GENERAL

An assault across a river normally begins with an attack to secure terrain on the exit bank. This may involve an air assault, but the bridgehead force normally conducts an assault by using pneumatic boats or by swimming amphibious vehicles.

The assault force normally crosses in waves, as sufficient boats are seldom available to carry the entire force across at once. It is a very complex operation, requiring synchronization and skilled application of technical procedures. Success requires training and extensive rehearsal.

Forces normally conduct an assault at night or during limited visibility due to the vulnerability of forces in small boats on open water. If an assault must be conducted during daylight, the assault site must be isolated by fires and smoke to reduce the force's vulnerability.

This chapter describes an assault-boat crossing. It focuses on conducting the crossing at night. It defines the organizational elements required to conduct an assault across a river and the necessary supporting techniques and procedures (see *Figure 8-1*).

TYPES OF ASSAULT CROSSINGS

Each lead battalion in a ground assault should have at least one ford or assault-boat site big enough to accommodate two companies abreast.

1. Engineers conduct far-shore reconnaissance.	13. The boat groups carry the boats to the river and launch the boats.
2. Assault and support forces conduct nearshore reconnaissance.	14. The company flotillas cross the river.
3. The assault force conducts a rehearsal (day).	15. The support force provides suppressive fires, if required.
4. The assault force conducts a rehearsal (night).	16. The assault force places smoke on the river, if required.
5. The assault force moves into the AA.	17. The assault force debarks, deploys, and attacks.
6. The company guides link up with the engineer boat platoons.	18. The second-wave force moves to the river.
7. The engineer boat platoons move into the attack position.	19. The boat groups return to the near shore.
8. The support force moves into support-by-fire positions.	20. The engineers mount the motors, if required.
9. The engineer boat platoons distribute and prepare the boats.	21. The second-wave force and cargo are loaded into the boats.
10. The company guides bring the assault force to the attack position.	22. The second-wave force crosses the river.
11. The boat teams man the boats.	23. The far-shore aid station is established.
12. The preparation teams prepare the far shore.	24. The assault force seizes the objectives.
	25. The assault force establishes the hasty defense.

Figure 8-1. Assault steps, summarized

Fording vehicles are more likely to be used in a hasty crossing than in a deliberate crossing because they allow the force to continue across the river without pausing to acquire other crossing means. A ford site should have 300 meters along the near bank at the entry point for deploying the support force.

RUBBER-BOAT CROSSING

The following factors should be considered when using rubber boats in a crossing. Rubber boats—

- Offer great opportunity for surprise in a silent-paddle crossing.

- Are a relatively fast means of crossing, especially when using outboard motors.

- Maneuver well in the water.

- Require limited, if any, entry-bank preparation—none is required on the exit bank.

- Require the separation of mechanized troops from their vehicles and equipment.

- Have limited carrying capacity, particularly AT weapons.

- Provide limited protection, mobility, firepower, and communications on the exit bank.

The unit protects itself during a rubber-boat crossing by moving silently, during periods of limited visibility, and crossing at a location where the enemy does not expect a crossing attempt.

Generally, an infantry platoon uses three rubber boats for its personnel and attached elements. If short of boats, the dismounted elements of an infantry platoon equipped with the M2 Bradley vehicle can fit in two boats. Allocating one squad per boat, when possible, preserves unit integrity.

For an assault using rubber boats, each company requires at least 200 meters along the river to disperse the boats and ideally 300 meters between companies. This is a total of 700 meters for a battalion assaulting with two companies abreast.

Control is very important, particularly at night when boats can easily become separated or lose their direction. Combat experience has demonstrated that engineer and infantry boat rehearsals before the crossing attempt are mandatory for success. The rehearsals should begin as soon as the unit receives the WO without waiting for the detailed crossing plan.

AIR-ASSAULT CROSSING

The following should be considered when using air-assault asset in a crossing. Air-assault assets—

- Require indirect approaches to avoid detection.

- Provide the element of surprise.

- Give greater flexibility to emplace personnel and equipment.

- Provide the rapid insertion of forces to the area where the enemy is located, if a LZ is available.

- Are greatly affected by weather conditions.

- Must be a high AD priority at the river, requiring suppression of enemy AD effort.

- Require the separation of mechanized troops from their vehicles and equipment.

- Are vulnerable to armored counterattacks and require a quick ground linkup.

Planning and execution are the same as for other air-assault operations (see *FM 90-4*). As with assault boats, rehearsals are necessary, particularly for troops not familiar with air-assault operations.

VEHICLE-SWIM CROSSING

Against little or no resistance, swimming the fighting vehicles may be practical in the assault stage. Swimming the fighting vehicles—

- Has minimal effect on troop organization and control.
- Provides troop protection, mobility, and firepower on the far bank.
- Provides early AT capability on the far bank by vehicle-mounted tube-launched, optically tracked, wire-guided (TOW) missiles.
- Reduces the number of vehicles to be rafted.
- Is a slow operation.
- Is considered risky because the vehicles maneuver poorly in the water and are extremely vulnerable to antiarmor weapons.

- Requires suitable entry and exit points.
- Requires vehicle preparation.
- Requires training in vehicle-swim operations.

Rapid reinforcement of dismounted assault troops with armored vehicles is so critical that it justifies using any expeditious method when swimming the first few fighting vehicles across. This includes winching, towing, or pushing the first ones across normally unsuitable places while engineers prepare better entry and exit points for the rest.

The space required to swim vehicles on line is 200 meters of front per company with 300 meters between companies. Less is required if they cross in a column. Commanders plan entry and exit sites to account for downstream drift when swimming the fighting vehicles.

ORGANIZATION

The specific organization used depends on METT-T factors, particularly the size of the bridgehead, the distance to exit-bank objectives, and the nature of the enemy's defense. Regardless of these factors, the assault battalion's TF organizes into support and assault forces and are assisted in the assault by other brigade units in support-by-fire positions.

SUPPORT FORCE

Each assault company has a support force under its control. This force covertly establishes a support-by-fire position along the friendly bank before the assault. It uses night vision and thermal sights to locate enemy positions. It also develops a fire plan to engage these positions and to provide suppressive fires on all suspected positions. When directed to engage, the support force destroys all known and suspected positions. The support force must

be positioned early enough to develop a detailed fire plan. The assault-force commander directs the support-force commander to lift or shift suppressive fires as necessary.

The support force normally consists of the tanks and infantry fighting vehicles of the dismounted infantry conducting the assault. If an attached light infantry battalion is conducting the assault, tripod-mounted heavy machine guns and AT missile systems (augmented by infantry fighting vehicles and tanks) provide supporting fires. The company XO controls these direct-fire weapon systems; however, the company commander gives the firing commands.

Supporting artillery battalions and mortar platoons provide indirect-fire support. The assault force has priority of fires from at least one artillery battalion during the assault. The artillery battalion does not normally fire a preparatory fire mission for covert assaults. The assault force assigns the batteries priority targets that they fire on upon request. This normally occurs after the initial wave is ashore or upon discovery. If the assault is not covert, the battalion fires preparatory fires that continue during the crossing of the first wave, lifting on command when the boats approach the exit bank.

Graphic fire-control measures are essential because of the danger of firing on friendly forces. Boundaries between companies should run along terrain features that are easily visible in the dark to help control indirect fire during the dismounted assault. Counterbattery fire is imperative to the success of the river crossing. The target-acquisition-battery-radar team deploys to cover the area before the assault crossing begins.

Smoke may not be used to support the first wave of a covert crossing because of the risk of losing surprise but should be used to hide later waves as they cross. If the crossing is opposed, a smoke haze should cover the first wave before it enters the water to reduce direct-fire effectiveness. The assault-force commander initiates smoke obscuration. If smoke generators are available, they are deployed to obscure a large length of the river. Additional smoke along multiple sites on the river conceals the true crossing area. This additional smoke may be from smoke pots if nothing else is available.

If units must fire smoke onto the far shore to cover the crossing area, they fire it on the command of the assault-force commander after surprise is lost. Mortars are the primary means of indirect-fire smoke. Direct-support artillery is generally reserved for supporting fires.

AD teams deploy along the near shore of the river to cover the crossing. Once in place, they remain until the brigade releases them. They can move across the river and link up with the assault force only after other SHORAD systems have taken position to cover the river. The crossing sites remain the priority AD area throughout the crossing.

ASSAULT FORCE

The first assault wave moves the bulk of the dismounted force across covertly. This force attempts to provide sufficient security on the far shore so that the second and later assault waves can cross after surprise is lost. The first assault wave carries—

- Rifle platoons.
- Attached assault engineers.
- Forward observers.
- The command group.

The organization of the first wave permits rapid deployment of the force into a tactical formation on the far shore. Individual boatloads retain unit integrity at the lowest level. The two basic boatload configurations are the rifle-squad boat and the rifle-platoon-headquarters boat (see *Figure 8-2*).

Each boat contains an engineer boat crew and a rifle squad. The squad boat also carries an engineer assault team, while the platoon boat carries the platoon headquarters. The boat-force commander is the senior occupant. He commands the force up to the attack position and after it debarks on the far shore. The coxswain is the "pilot in command" and commands the force from the point that it mans the boat in the attack position until it debarks on the far shore.

First-wave boats carry only critical cargo, such as AT weapons, machine-gun ammunition, demolitions, and engineer tools that are required for reducing obstacles.

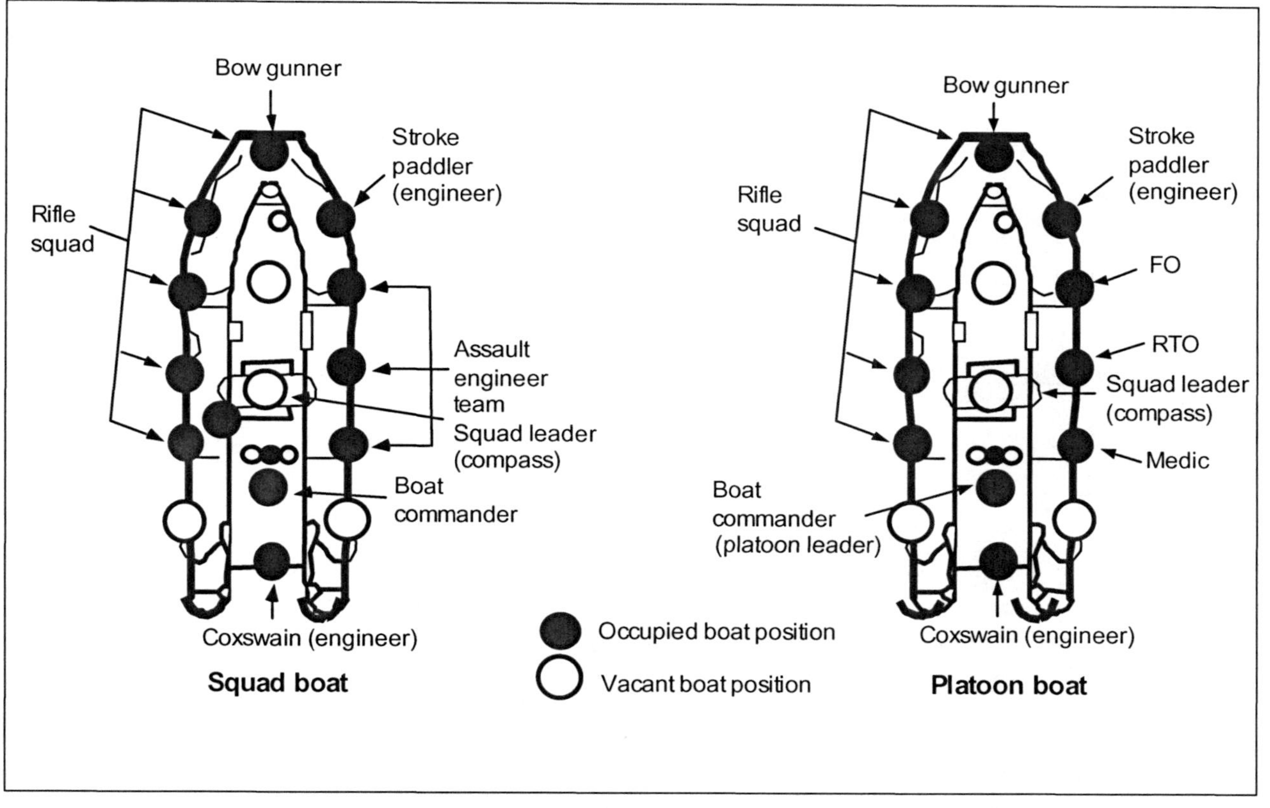

Figure 8-2. Boat-load configurations

Platoon boats form a boat group of three boats that are spaced 20 meters apart on the water. The boat group forms into a "V," with the platoon leader's boat acting as the guide boat in the center. The two engineer assault teams are from an engineer squad, with a squad leader commanding the team in the right boat and an assistant squad leader commanding the team in the left boat. The assault teams re-form into a squad upon debarking.

Platoon boat groups form into company flotillas (see *Figure 8-3, page 8-6*). The company commander commands the guide boat in the center platoon. The company command group disperses between boats, filling in vacant boat positions. Platoon guide boats maintain a 40-meter interval (two-boat interval) between boat groups.

The first wave of the assault may consist of three company flotillas crossing on line. Battalions do not have a prescribed crossing formation. Each company crosses in its own zone and attacks it own objectives.

All undamaged boats return to the near shore after carrying the first wave. The senior coxswain of the group will consolidate the boats and stroke paddlers into one (or more) boat and daisy-chain the other boats to the lead boat to expedite the time required for the boats to return to the near shore. The second and later waves carry across the remaining troops and materials that are necessary to seize the far-shore objective.

The second wave carries company aid stations and may include the battalion command group. Since sufficient AD systems

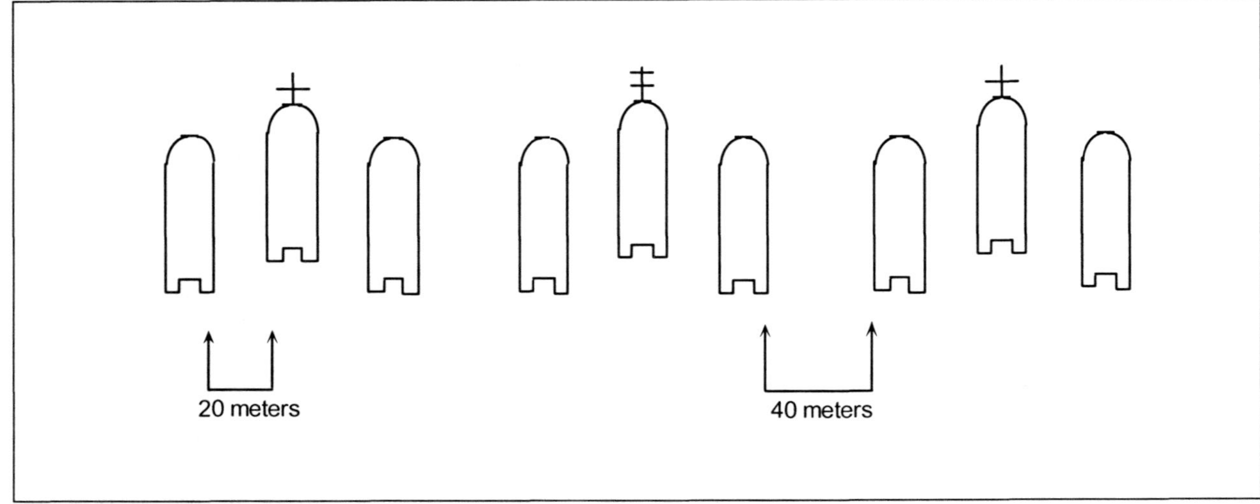

Figure 8-3. Company flotilla

are in place to cover the crossing area, the brigade may release some or all of the battalion AD teams to cross in the second wave.

The second wave also transports additional material and ammunition that is not required for the initial assault but necessary to establish a defense. This may include antiarmor weapons, mortars, ammunition, laser designators, mines, or pioneer tools. It normally includes tripod-mounted weapons, such as M2HB .50-caliber machine guns, TOW AT systems, Mark 19 40-millimeter grenade launchers, and the ground-laser location designator (GLLD).

If secrecy is not required for the second wave because the first wave is in combat, or if the enemy has begun to fire on the crossing area, then outboard engines are used to propel the boats so that paddlers are not necessary.

The immediate movement of some heavy AT weapons across to support the dismounted assault battalion is essential. This is critical enough to justify extraordinary actions. As vehicles carry all heavy AT weapons, engineers concentrate on forcing a few critical vehicles carrying heavy weapons across immediately after the second wave. They hand carry heavy weapons, if necessary, even before direct fire and observed indirect fire has been removed from the crossing area. Vehicles cross by swimming or fording or are dragged or rafted across.

CAEs begin bank preparations on both the near and far shore, using hand tools and equipment where possible. They swim an M9 ACE or deep ford a bulldozer to get a winch capability to the far shore. Bradley vehicles either swim or ford, with towing assistance if necessary. A BEB can tow Bradleys if the current's velocity is too high. Using a block and tackle fastened to a tree or picket holdfast, a BEB can help Bradleys leave the water over unprepared banks. If high-mobility multiwheeled vehicle (HMMWV) weapon carriers are available, they can be waterproofed and pulled across on the bottom with a winch cable. If absolutely necessary, rafting can be used, but this risks destroying equipment that will be critical later in the crossing.

ENGINEERS

Engineers supporting the assault are attached to the assault unit as described in previous paragraphs. Each assault company receives an engineer platoon that accompanies the assault force to its objective, helping it fight through obstacles and prepared defenses. The engineers help the assault force establish hasty defenses after it has seized its objectives. Engineers normally come from the division engineer battalion that supports the brigade.

Boat engineers—

- Operate the boats and cross the assault force.

- Are in direct support of the assault battalion until it has secured its objectives.

- Remain on the water after the assault force has crossed and continue to carry men and materials across in assault boats until heavy rafts can take over the mission.

- Improve exit and entrance banks for rafts and boats and assist with crossing the initial heavy weapons.

- Come from the engineer battalion that will remain on the river operating the crossing area.

Two boat engineers are assigned to each assault boat. They are the coxswain and the lead paddler on the right side of the boat (stroke paddler). The stroke paddler controls the stroke cadence during the assault crossing. The boat engineers paddle the boats back for the next wave. Outboard motors normally are used during the second wave.

Normally, an engineer platoon must operate the boats for a first-wave assault company. An engineer company can cross the assault battalion of a brigade. Each assault company requires 9 boats plus a safety boat. The assault battalion requires 30 boats to carry the assault companies plus 1 for the battalion commander. If less are available, some companies may not cross in the first wave.

PREPARATION PHASE OF THE OPERATION

Similar to conducting a deliberate breach of a complex obstacle, river-crossing operations require intelligence collection, detailed planning, and preparation. These must all be synchronized to allow the force to maintain its momentum and surprise the defender at the point of penetration.

FAR-SHORE RECONNAISSANCE

Tactical reconnaissance of the far shore must cover a broad front to a significant depth to determine the details of the terrain and the enemy's defenses. This should occur early and cover sufficient terrain to disguise the actual crossing area.

Engineers conduct a technical reconnaissance of the far shore, focusing on the immediate crossing area. An engineer light diving team conducts a reconnaissance at night. If a diving team is unavailable, then a swimming reconnaissance team is made up from the engineer unit supporting the crossing. Strong swimmers (Red Cross-certified lifeguards or water-safety instructors) from the engineers supporting the crossing make up the reconnaissance party if divers are not available. Two swimmers make up a reconnaissance team to scout a company crossing area.

The reconnaissance team carries heavily lubricated weapons and wears LBE. They wear subdued face masks and running shoes and use swimming fins. Divers must wear Class 5 life jackets as flotation devices (US Army flat foam-filled life jackets will not serve). The divers camouflage their faces and hands and tow any necessary equipment in bundles.

Divers must carefully avoid splashing. If necessary, they wear weights to ensure that kick strokes are underwater. The party enters the water far upstream from the actual crossing site and floats with the current while crossing. Divers use the sidestroke, facing each other and observing behind the other diver. This allows 360-degree observation and communication by hand and arm signals. When the divers approach the shore, they switch to the breaststroke so that they can observe the landing area. Divers must use stealth and caution when approaching the beach. They must keep a low profile in the water and also on the beach.

When the divers reach shallow enough water and determine that the situation is safe for landing, they remove their fins. If they can immediately enter the woods upon leaving the water, they do so in a rush. If the woods are a distance from the water, one diver remains in the water just at the waterline and covers the other as he moves quickly across the beach. Once the inland diver has reached the edge of the woods, he covers his partner, who is moving across the beach to the same position.

Critical information requirements include—

- The characteristics of the bank at assault-boat landing areas.

- The depth of the water to a distance of 4.6 meters off shore.

- Any obstacles along the shore.

- The locations of enemy observation posts.

The reconnaissance team checks potential areas identified from the near shore and evaluates each based on its ability to support assault boats and disembark troops. The reconnaissance party also checks areas where raft and bridge centerlines can be installed.

Far-shore reconnaissance is conducted early and at multiple sites along the shore to generate information necessary for planning and selecting the most suitable areas. Maneuver units, with support from the engineers, conduct far-shore reconnaissance.

FAR-SHORE PREPARATION

The far shore is prepared immediately before the assault crossing. The preparation team consists of a two-man reconnaissance team and a two-man cargo team with an inflatable reconnaissance boat; both teams are from the supporting engineers. The reconnaissance team that conducted the far-shore reconnaissance is normally best-suited to do the far-shore preparation. The preparation team installs landing markers for the flotillas. A separate team normally marks each company zone to speed up preparation.

The reconnaissance team and the cargo team are equipped the same as the reconnaissance party and use the same techniques. The reconnaissance team crosses first, floating downstream to the landing site with the current. Upon landing, they move to the correct landing site for the assault landing and signal for the cargo team to cross. The reconnaissance team installs transit lights to guide the cargo team as it crosses.

Signaling is accomplished by sending a prearranged Morse Code letter using a flashlight equipped with an opaque filter. The transit lights consist of either two flashlights with opaque filters and directional cones or two chemical lights in their foil wrappers with small areas torn open to release light. The team installs the lights so that one is about 1 meter above the water and the other is about 2 meters above the water and 2 meters behind it, facing 45 degrees upstream.

The cargo team waits until signaled to cross. It uses a three-man reconnaissance boat as a flotation device to carry marking materials, mine detectors, night-vision goggles, and a radio. The reconnaissance boat is covered with a camouflage net section and is partially deflated after loading so that it floats low in the water to reduce its signature. The camouflage net is secured to the lifelines to aid in holding the cargo in the partially submerged boat. The cargo team crosses oriented on and swimming slightly upstream of the transit lights so that it can drift into the shore with the current, limiting the noise and the splash.

The preparation team installs landing markers as its first priority. These are the same types of markers used to guide the cargo team. They must be adequately visible to the assault force but dim enough not to harm night vision. If flashlights are available, they have opaque and/or colored filters installed to limit the light output. Chemical lights remain in the foil wrappers with only enough foil removed to provide the necessary light. All landing markers are transit lights that mark the position and help the boats set the proper course relative to the

current. Normally, if the current is less than 0.5 MPS, the lights are set perpendicular to the river. If the current exceeds 0.5 MPS, the lights are set at a 45-degree angle to the river, facing upstream. Double transit lights mark the center boat group's landing area, and single transit lights mark the flank group's. If colored lights are available, green lights mark the right boat group's landing area, white the center's, and red the left's (see *Figure 8-4*).

The preparation team also makes a final examination of the landing areas for mines or obstacles. If it discovers isolated mines, it marks them and the routes around them. If the team finds a major minefield that will significantly hinder the landing at a site, it either notifies the assault force and moves the site upstream or downstream to avoid the mines or attempts to reduce the minefield. Once the preparation is complete, the team signals the assault force to begin crossing, initiating the movement of the first wave carrying the boats from the attack position. The preparation team then finds cover near the landing area for the center boat of a predesignated boat group (generally the center boat group) and awaits its arrival. This boat group is especially alert for linkup with the preparation team. While waiting, the team

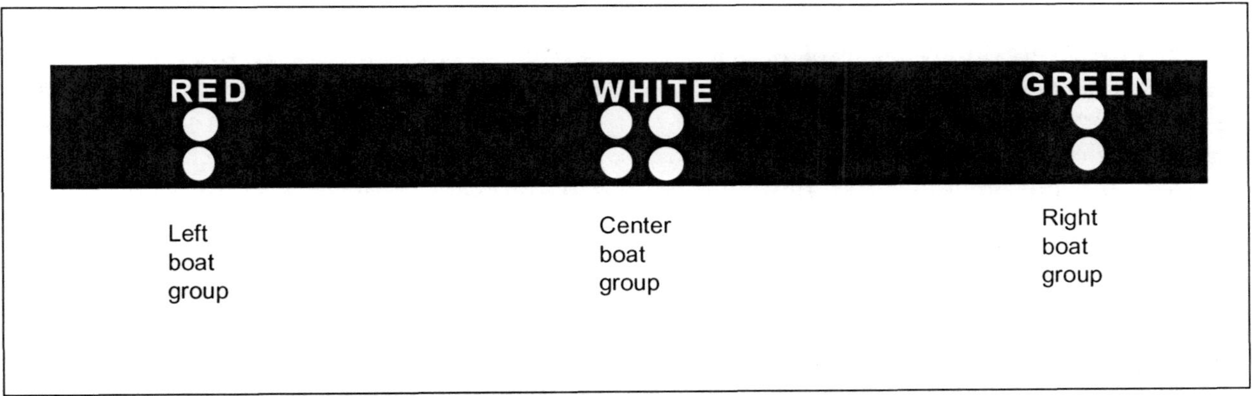

Figure 8-4. Landing marker lights

continues to watch for enemy activity and alerts the assault force of any significant changes.

NEARSHORE RECONNAISSANCE

Units must be extremely careful to hide reconnaissance elements conducting near-shore reconnaissance in the crossing area or to deceive the enemy about what they are doing.

Battalion and company command groups must conduct a daylight reconnaissance of the crossing area. They must see the embarkation and debarkation points and key landmarks to help guide the force when crossing. They must also see the attack position and the routes from it to the river. Company guides must walk the routes from the dismount points to the boat-group positions within the company's attack position. Engineer boat coxswains must see the routes they will traverse from the attack position to the water.

Support-force leaders and vehicle command-ers must covertly select firing positions and locate concealed routes into the positions for their vehicles during daylight. They should identify sectors of fire and conduct extensive observation within the sectors to acquire specific targets.

ASSAULT-FORCE REHEARSAL

An assault-boat crossing cannot be conducted effectively in the face of opposition without thorough rehearsal. If possible, the force should conduct two rehearsals. One should be during daylight to learn the procedures, and one should be at night under actual assault conditions.

The rehearsal area should be similar to the actual crossing area but away from the river to preserve secrecy. Generally, a rear-area river is the rehearsal area.

Before rehearsal, the boat crews and infantry train together in the actual boat teams assigned for the crossing. Soldiers receive their boat assignments and practice in their assigned positions until the boats can move effectively on the water. The training must include carrying and launching the boat, embarking, watermanship, emergency actions, debarking, and hasty-defense preparations.

NOTE: After the rehearsal, boat assignments must not be changed.

During training, the coxswain forms the boat team. He forms the crew members in a column of twos in the relative positions that they will occupy in the boat, with passengers at the rear of the two columns. He then numbers the crew. The right-side paddlers are 1, 3, 5, and 7, and the left-side paddlers are 2, 4, 6, and 8 (both sides from bow to stern). The stroke paddler is always number 1, and the coxswain is always number 15, regardless of the number of paddlers used. Passengers are numbered consecutively from bow to stern starting with number 11, who is always the bow gunner. The coxswain addresses all crew members by number. When the coxswain wishes to address a command to a pair of paddlers, he uses their numbers together, as in "1 and 2" and "3 and 4."

Figure 8-5 shows only 8 paddlers. The boat can carry 15 soldiers. If fully loaded, the boat requires 10 paddlers. Boat-position numbers do not change.

All forces participating in the assault crossing rehearse together. The support force moves into position, and the assault force crosses in the same waves it will use for the actual crossing. The rehearsal should cover the AA through to the seizure of the assault-force objectives.

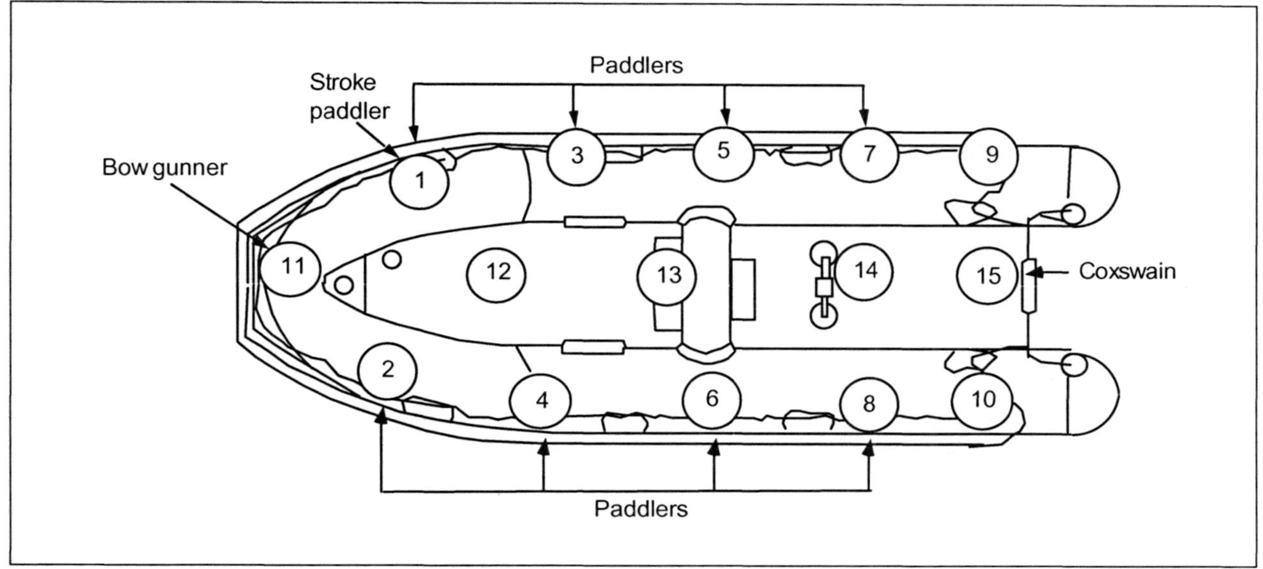

Figure 8-5. Boat numbering

EXECUTION PHASE OF THE OPERATION

The objective of an assault river crossing is to project combat firepower to the exit bank without being detected by the enemy or, once detected, project it at a faster rate than the enemy can concentrate forces for a counterattack. The use of air assets is desired; however, there are normally not enough assets available, or the risk of being detected is to great. To maintain momentum and allow maximum combat power across quickly, the maneuver force negotiates the river on a broad front. Detailed planning and specific responsibilities allow units to cross and quickly establish a tactical foothold on the far side to—

- Prevent the enemy from indirect-fire observation.

- Allow rafting and bridging operations to begin.

ATTACK-POSITION PROCEDURES

The attack positions must be large enough to accept a dismounted infantry rifle company. They should—

- Be accessible to trucks or carrying parties bringing the assault boats.

- Be concealed from hostile ground and aerial observation.

- Be connected with clearly defined foot routes to the river.

- Be within 100 to 200 meters of the river.

- Be in defilade from hostile flat-trajectory fire.

Trucks carry assault boats and life jackets as far forward as possible without compromising secrecy. They are met at the designated unload position by the engineer platoon and company guides from each attack position, who will unload the truck and carry the boats into place. The platoon can carry two at a time, so this will require five trips. If possible, HMMWVs moving at a low speed to minimize noise can carry several boats at a time into the attack position.

Within the attack position, boat crews disperse assault boats and life jackets along the boat-group routes to the river. The safety boat is positioned as the last boat in the downstream boat group. The remaining life jackets for passengers and the coxswain are arrayed behind the boat.

After the boats are prepared, each engineer squad provides a guide to bring each platoon from its AA to the nearshore crossing site. The platoon leader sends the guide party to the AA, where each guide links up with his boat group. The remaining engineers establish local security around the attack position and await the boat groups.

Soldiers arrive in the attack position with their weapons cocked on an empty chamber, selector switch on *SAFE*, and magazines removed. Squad leaders must verify this in the AA before moving to the attack position. The soldiers are organized, without the boat engineers, into boat teams and boat groups in the AA. They travel as boat groups. When they arrive at the attack position, their guide leads them directly to their boats.

When the boat team arrives at its boat, the coxswain commands, "*Crew, boat stations.*" Each team member takes his proper boat position, with passengers lining up to the rear. The coxswain then directs the team to load and check weapons. The team inserts the rifles' magazines and verifies that they are seated. However, the team does not chamber the rounds. All weapons remain on *SAFE*. Squad leaders verify that all weapons are on *SAFE*. The coxswain then directs the team members to sling their weapons and don their life jackets. Paddlers sling their rifles diagonally so that the barrel extends up over the shoulder which will be away from the boat when standing alongside and facing forward. Odd-numbered paddlers sling their rifles over their right shoulder, even numbered over their left. This allows carrying the boat at high carry and reduces interference with paddling. Muzzles must be

up during all boat operations to prevent punctures. The teams then await the command to proceed to the water.

EMBARKING PROCEDURES

On order of the company commander, the paddlers of the boat crew carry the boats to the river. They make no unnecessary stops from the time of departure from the attack position until the boat reaches the bank. The coxswain directs either "*Low carry*" or "*High carry.*" In low carry, crew members lift the boat to about knee height, by the carrying handles while facing forward, and carry the boat at arms length. In high carry, crew members lift the boat to about head height, place it on their inboard shoulders, and carry it while gripping the carrying handles with their outboard hands. Normally, high carry is used for long distances, and the boat is shifted to low carry when approaching the bank. Paddles remain in the boat during carry procedures. Remaining crew members follow the boat to the water.

The boat crew may launch the boat either bow first or stern first; however, bow first is the preferred method. The boat is launched—

- Bow first whenever the water is shallow enough for the team to wade in carrying the boat at low carry.

- Stern first when the water is too deep for wading or when the launch point has steeply sloped banks.

Bow-First Method

On the coxswain's command, "*Launch boat,*" team members perform a low carry and move into the water at a fast walk. When the depth of the water is such that the boat floats free of the bottom, all hands continue pushing it into the river, remaining at their relative positions alongside the boat.

As the water reaches the knees of the first pair of paddlers, the coxswain commands, "*One and two in.*" The first pair of paddlers climb into the boat, unstow their paddles, and give way together. The coxswain orders each pair of paddlers into the boat in succession by commanding, "*Three and four in,*" "*Five and six in,*" and "*Seven and eight in.*" The pairs climb into the boat on command, break out their paddles, and pick up the stroke of the stroke paddler.

The coxswain orders the passengers into the boat after the paddlers by commanding, "*Eleven in,*" "*Twelve in,*" and so forth. Passengers board over the stern and move forward in the boat to their positions. The coxswain enters the boat last and sounds off, "*Coxswain in, hold water.*"

Stern-First Method

On the coxswain's command, "*Launch boat,*" team members perform a low carry and carry the boat stern (rear) first to the water's edge. They launch the boat by passing it back along the line of team members. When the stroke paddler can no longer help pass the boat back, he moves to the bow of the boat and handles the towing bridle. Other team members follow suit, taking their places along the towing bridle between the stroke paddler and the boat.

When the boat is in the water, the coxswain enters the boat and takes his station. He orders the boat team to load, starting with the rearmost left-hand paddler, by commanding *"Eight in," "Seven in," "Six in," "Five in," "Four in," "Three in,"* and *"Two in."* Passengers embark next as he commands, *"Fourteen in"* and *"Eleven in."* When the coxswain is ready to cast off, he allows the boat to drift back and turns it to face across the river.

If motors are to be mounted before the first-wave crossing, the coxswain brings the boat in to shore stern first after the boat is manned and holds it in place either by a line to shore or by holding bottom. Two engineers wade to the boat carrying its motor and mount it on the transom.

TACTICAL CONTROL AFLOAT

The coxswain navigates the boat and directs the paddlers. He controls the movement of the boat in the water as well as embarkation and debarkation from it. He ensures that the guide boat maintains the proper station. The boat commander sits in front of the coxswain and directs the boat in an emergency. He also commands the boat occupants upon landing until the unit has re-formed. The boat commander directs fires from the boat, if necessary.

Each platoon has a platoon guide boat, which contains the platoon headquarters. Other platoon boats position themselves to either side of the platoon guide boat as wingmen to maintain a 20-meter interval for protection against fires and to allow dispersion on landing. They follow the guide boat and land when it does. They open fire from the boat when the guide boat does.

Each company has a C^2 boat, which carries the company commander and leads his flotilla. Platoon guide boats position themselves at double-boat intervals from the C^2 boat, maintaining a 40-meter spacing between boat groups. The C^2 boat is normally the lead boat of the center platoon.

The battalion command group remains on the near shore until the assault wave has landed. The commander controls the nearshore direct fires and directs changes in landing points if elements of the first wave encounter difficulties. He also directs changes for the following wave. The commander has his own boat and crosses on his own schedule, but he normally crosses with the second wave.

The command group normally does not cross in a single boat but is distributed among several boats.

Guide boats in all boat groups are responsible for ensuring that their group lands at the proper place. Landing marker lights are installed as transit lights to assist navigation on the water. The coxswain will see two lights, one above the other. If the boat is moving straight to the landing, the lights will be straight in vertical alignment. If not, the lower light points in the direction the boat must go to be exactly headed for the landing. The boat will not head directly for the transit lights except when the river has no current. The boat heads for the far shore so that the boat's true course is directly for the lights (see *Figure 8-6*).

Normally, the boats will cross slightly upstream from the landing so that they can drift in with the current. To do this, they align so that the lower transit light points slightly upstream.

If the force is conducting a crossing where smoke is necessary on the water and it obscures the far shore, other navigation methods it could use include stringing ferry lines across the river for the boats to follow, using floating markers, or traveling on a compass heading.

WATERMANSHIP

Watermanship includes all the skills that the boat crew must exhibit to properly control its boat in the water. It includes individual paddling skills, responsiveness to commands, and the skill of the coxswain.

Individual paddlers use a paddling technique where they push the paddle vertically into the water, roughly 1 meter to their front, and then power it back through the water by pushing with the upper hand while using the lower (guide) hand for control. At the end of the power stroke, they remove the paddle from the water, turn it outboard and parallel with the water's surface (feathering), and move it forward for the next stroke. The stroke is silent, with the paddlers careful not to strike the side of the boat or to splash.

The stroke paddler sets the pace to control the paddlers. He receives oral commands from the coxswain and establishes and maintains the paddling pace. All paddlers match the stroke of the paddler in front of them except for the number two man, who matches his stroke with the stroke paddler. If the boat crew has difficulty paddling in unison, the coxswain can exercise oral control by calling cadence. The normal paddling speed is 10 strokes per minute for stealth and 30 strokes per minute for speed.

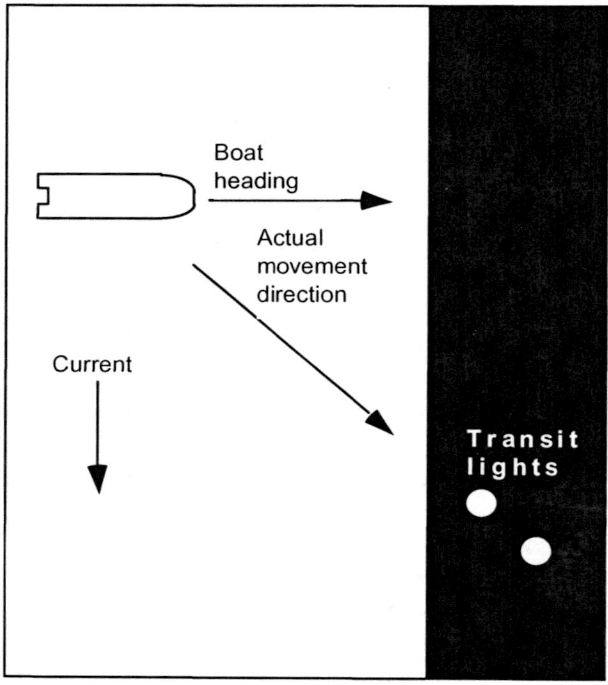

Figure 8-6. Boat course

The coxswain uses the following commands to control the boat:

- "*Hold water.*" Paddlers hold their paddles motionless in the water with the blade perpendicular to the direction of motion.

- "*Give way together.*" Paddlers stroke in unison following the rhythm set by the stroke paddler.

- "*Slow stroke.*" The stroke paddler paddles 10 strokes per minute.

- "*Fast stroke.*" The stroke paddler paddles 30 strokes per minute.

- "*Backwater.*" Paddlers paddle backward in unison with the stroke paddler.

- "*Rest paddles.*" Paddlers rest their paddles across their legs.

- "*Hold bottom.*" Paddlers thrust their paddles straight down into the river's bottom and hold them against the side of the boat as a temporary anchor.

- "*Land boat.*" The stroke increases to 30 strokes per minute, with each paddler digging deep into the water for power to drive the boat up on shore. The stroke paddler stows his paddle as soon as the boat grounds, then disembarks and secures the towing bridle to the shore.

- "*Right, backwater, left, give way together.*" When paddlers execute these commands, the boat turns rapidly to the right. When the boat has turned to the new desired course, the coxswain commands, "*All, give way together.*"

- "*Left, backwater, right, give way together.*" When paddlers execute these commands, the boat turns rapidly to the left. When the boat has turned to the new desired course, the coxswain commands, "*All, give way together.*"

The coxswain can make minor adjustments in the boat's speed by directing, "*Slow the stroke*" or "*Speed the stroke.*"

The coxswain must take the current's velocity into account when trying to hold a course. In low-velocity current, the boat can travel a relatively straight course across the river by crabbing slightly upstream. To do this, the coxswain aims the bow of the boat slightly upstream while sighting on the land mark. If the mark remains on a constant bearing (it does not drift upstream or downstream), the boat is crabbing correctly and is headed directly for the landing.

If the current's velocity is too high for successful crabbing (over 0.5 MPS), either the boat must start upstream or the coxswain must steer a figure-eight pattern. In both cases, the boat should approach the landing heading into the current to avoid the danger of broaching. If the boat is launched from far upstream, it generally follows a course similar to the dotted course in *Figure 8-7, page 8-16.* If the coxswain follows a figure-eight course, he steers upstream until aligned with the transit lights, then lets the bow drop downstream and guides by using the lights until he reaches the landing point. He then steers upstream to the landing marks (see *Figure 8-7,* solid line). These techniques minimize the amount of time the boat will be traveling slowly against the current while near the enemy shore.

The need for a figure-eight course is determined during reconnaissance. The flotilla command boat sets the figure-eight course, completing the downstream turn in alignment with the transit lights. Remaining boats simply maintain station until the last turn upstream toward the landing area. Boat groups then head directly for the transit lights.

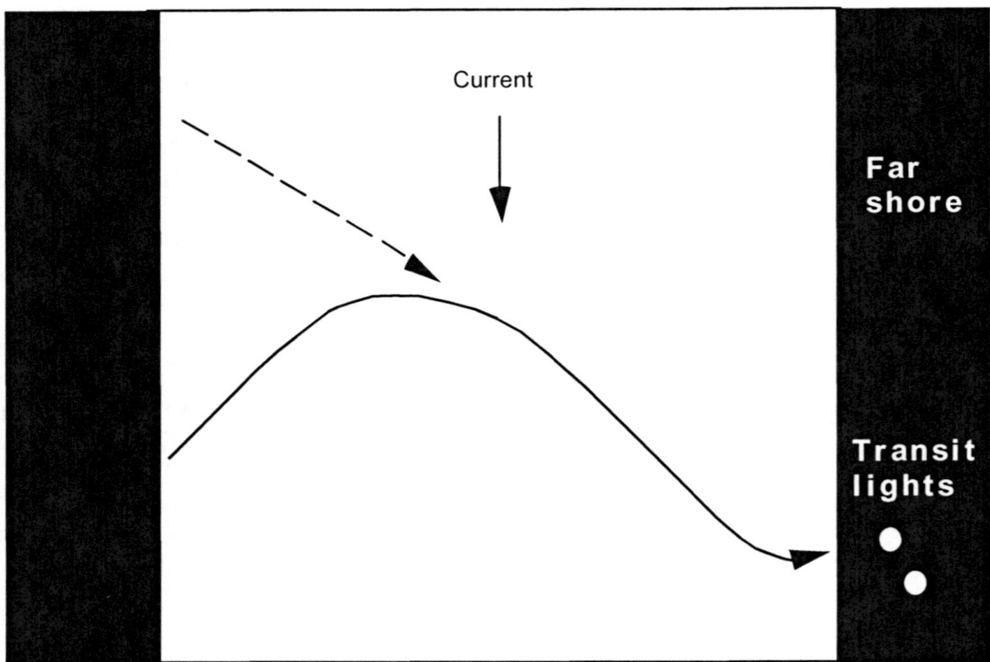

Figure 8-7. Figure-eight course

Eddy currents (eddies) occur at channel bends, near points of land, and at places where the bottom is uneven. Eddies can be dangerous to small boats. The coxswain must be alert for them.

OBSCURING WITH SMOKE

The purpose of smoking the crossing site is to achieve a haze over the water that can render direct and indirect fires less effective. Smoke may be used during river-crossing operations to—

- Conceal the movement of the initial assault force.

- Isolate the exit bank of the river for rapid occupation by maneuver forces.

- Conceal emplacement of crossing means, such as engineer bridges.

During river-crossing operations, smoke may be used for—

- Obscuration

- Screening.

- Marking.

- Deception.

It is particularly important not to produce a column of smoke above the water that can pinpoint the crossing location. For this reason, smoke is not used if conditions will not hold it close to the surface.

Smoke production depends on wind direction. If the wind is blowing from the near shore toward the far shore, smoke generators or support-force vehicles can effectively smoke the crossing. If the wind tends to blow parallel to the river, nearshore smoke should not be used, as it will make a smoke wall that will silhouette boats on the river. In this case, floating smoke pots anchored across the width of the river can produce effective smoke. If the wind is blowing from the far shore to the near shore, smoke pots or mortar smoke on the far shore can be effective.

DIRECT-FIRE REACTION

If the boat is subjected to heavy direct fire while crossing, the boat commander may direct all personnel, except the bow gunner, to stow their paddles, slip over the side while holding the safety line, and propel the boat to shore by kicking with their feet.

All boats have a designated gunner at the bow that is armed with either a squad automatic weapon (SAW) or a bipod-mounted machine gun. The gunners do not fire unless the boat commander orders them to. If ordered to fire, the gunners engage the most dangerous target or suppress the landing area. More often, the gunners engage enemy weapons firing on the assault force by firing back up the line of enemy tracers. If two passengers are available to be boat gunners, the second back from the bow should be armed with a grenade launcher.

To preserve their night vision, all paddlers observe the paddle of the man to their front. They do not look at the enemy shore from where the muzzle flashes are coming.

INDIRECT-FIRE REACTION

If the boat is subjected to heavy artillery fire while crossing and the boat commander directs, the coxswain turns the boat downstream and propels it at a fast stroke with the current out of the artillery impact area. If the boat is equipped with a motor, it is started and the paddlers stow their paddles and maintain a low posture.

DEBARKING PROCEDURES

The manner in which the coxswain orders the boat team to land the boat depends on the depth of the water at the landing point.

Shallow Water

As the boat nears the landing point, the coxswain directs the boat toward the landing and orders, "*Land boat*." As the boat grounds, paddlers stow paddles and disembark over the side into the water. They then hold the boat for the passengers to disembark. The stroke paddler secures the boat and awaits to return it.

Deep Water

As the boat comes along the shore, the coxswain orders, "*Stroke out*." The stroke paddler stows his paddle and, with towing bridle in hand, gets out of the boat onto the shore. He then pulls the boat up close to the shore and secures it if he can. Otherwise, crew members will have difficulty debarking. The other crew members stow their paddles. The coxswain then directs debarking by number, beginning with the passengers, then the shoreside paddlers, and finally the riverside paddlers. The coxswain is the last to leave the boat. He and the stroke paddler secure the boat and await to return it.

Immediately upon leaving the boat, the boat team forms a hasty perimeter. The bow gunner moves directly forward, roughly 10 meters, and drops prone, observing to his front. The left-side squad members move up and form a prone semicircle to his left. The squad leader takes charge of his squad and directs all soldiers to drop their life jackets. He then awaits orders from his platoon leader.

BOAT RETURN

As soon as the boat team has formed a hasty perimeter and dropped their life jackets, the stroke paddler recovers them and returns them to the boat.

The boat engineer squad leader (the senior engineer with the boat group) takes charge of all three boats in the boat group. He supervises the tying off of all three boats in a trail and loads all six engineers into the front boat. They then paddle the boat back to the friendly shore, towing the other two boats (see *Figure 8-8*).

On the return, the boat group travels in a relatively straight line to gain distance from the enemy shore as rapidly as possible. This will cause the group to drift downstream. Upon reaching the near shore, the boat group turns upstream and travels close inshore until it reaches its original departure point (see *Figure 8-9*). A guide from the engineer platoon headquarters guides them in for the next wave.

If the boats have outboard motors, all three boat crews start their motors on command of the boat engineer squad leader and return independently to the near shore.

MOTOR PROCEDURES

If motors are available, they speed the crossing significantly. Normally, the first wave uses paddles to cross covertly. After the boats return from carrying the first wave, the motors are mounted. If the boats can be placed in the water without enemy observation (in a lagoon or barge basin, for example), the motors are mounted on the boats before the first wave crosses. In this situation, the motors can be started immediately if the crossing is discovered. The motors are also available for returning the boats after the first wave.

If a covert crossing cannot be achieved, the first wave may cross the river powered by motors. In this case, the motor is mounted before the boat crew and passengers carry the boat to the shore. Two additional engineers are provided to help carry the stern of the boat to the shore. The crew paddles the boat while the coxswain starts the motor in order to reduce exposure time on the river. This technique must be practiced during the rehearsal.

If time permits or the distance to the water is great, the two-man team of engineers from the crossing-area engineer battalion carries the motor to the water and mounts it on the boat. The boat is manned and held with the bow toward the river and the stern to the shore. If the bottom is shallow, the paddlers hold bottom. If the water is too deep or the current too strong, a line is fastened to the boat stern to hold it against the shore. The mounting team wades out to place the motor on the stern and fastens it in place. The coxswain directs the paddlers to give way together after the motor is mounted. He then starts the motor with the

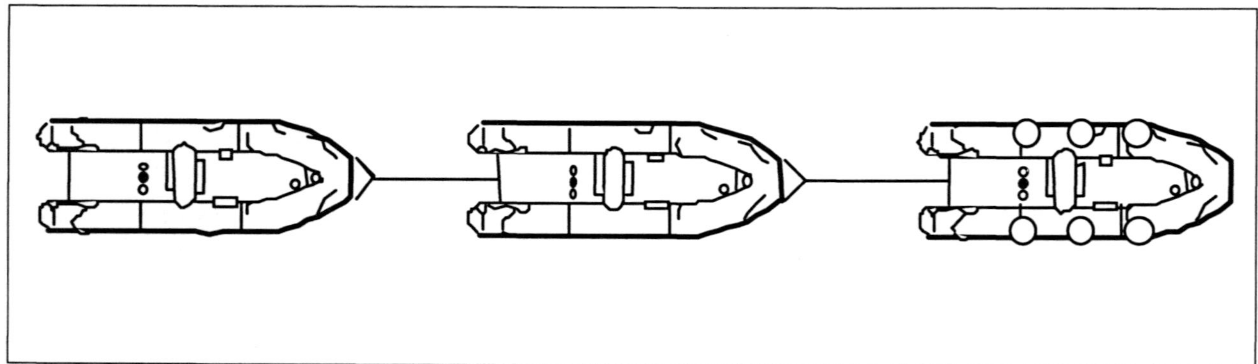

Figure 8-8. Boat return

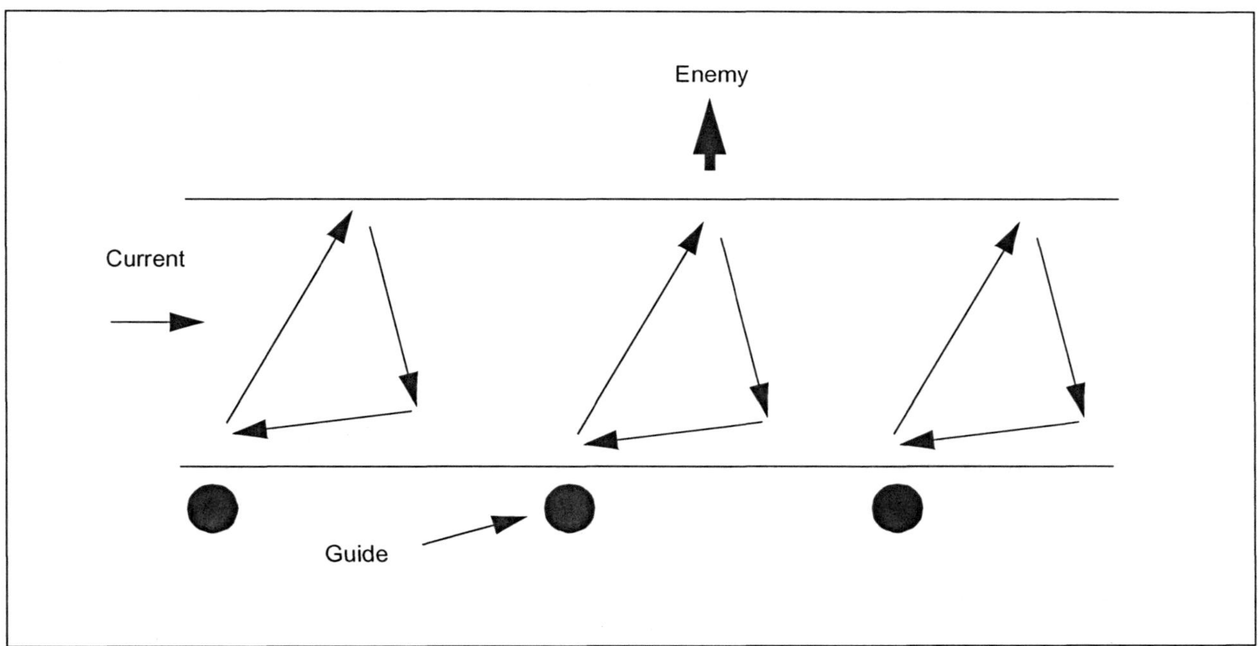

Figure 8-9. Operation of boat group

boat under way. If the boat has too few occupants to move effectively by paddles (during the second wave, for example), the boat remains at the shore until the coxswain starts the motor.

Preparation is critical for success with outboard motors. The primary problem is hard starting. All motors are started and run up to operating temperature during preparation. If any are difficult to start, replacement motors are substituted (the hard-starting motors become backups). After mechanical checks and warm-ups, the fuel tanks are completely filled with the correct fuel and oil mixture to eliminate condensation. In cool or cold weather, the motors are kept warm until needed, using a warming tent, ambulances with medical markings covered, a heated building, sealed wrapping, or other means.

CARGO PROCEDURES

Porters detailed from the assault force bring the cargo forward. They carry it to the waterline at the boat launch point to await the return of the boats. When the boats return, the porters load and secure the cargo to the boat. If the cargo includes heavy or pointed items, a temporary plywood floor is placed in the boat before loading.

Porters accompany the cargo to the far shore to unload it. The cargo is unloaded into caches until carrying parties are sent back from the assault force to get them.

CASUALTY PROCEDURES

Platoon medics accompany assault forces in the first wave. They carry their medical bags and night-vision goggles but do not have litters. They treat wounded where they fall, sending walking wounded back to the landing area and leaving more severely wounded where they were treated.

The second wave carries senior aidmen with equipment to establish a far-shore casualty collection point in each company zone. The aid station should provide a blackout shelter, such as a tarpaulin or small tent, for patient examination along with emergency medical supplies and quantities of intravenous fluids. The second wave also carries litter teams formed from the headquarters elements of the assault force. The litter teams carry wounded back to the collection point. The senior aidman at the collection point performs triage and treats patients. Priority patients are evacuated by assault boats as they become available. All other patients wait until rafts are available.

SAFETY

Safety is as important in combat as it is in peacetime training. Procedures are established and soldiers are trained in peacetime to be safe in combat. Loss of a soldier to an accident in combat is just as intolerable as losing a soldier in peacetime and is potentially far more dangerous to the force. Safety procedures are particularly important when considering the risks during assault river crossings, where the lost soldier may be the key to mission success. Therefore, all safety procedures must be followed in combat.

The most important safety procedure is building a well-trained force. Nothing is a greater safety risk than allowing a force of untrained soldiers to undertake a complex, potentially hazardous task where the well-being of all depends on each soldier knowing his job. Peacetime training should never be avoided because of the potential hazards of a necessary combat task. Training to standard in a controlled environment is the only way to surmount the hazards and build confidence in the soldiers' ability to accomplish their mission.

Life jackets are always worn when using assault boats. If Class 5 life jackets (German-army style) are available, they are worn over LBE and the diagonally slung rifle. The Class 5 life jacket will support a soldier so equipped and hold his head out of the water. If a life jacket providing lesser flotation is used, such as the standard US Army flat foam-filled life jacket, it is worn over the uniform. The LBE is worn over the life jacket, with the belt unfastened and the rifle slung diagonally over all. Rifle slings are turned around so that the free end is always away from the weapon. This allows rapid jettison of the rifle in the water by pulling the free end of the sling to release the fastener.

Weapons are always carried in the boats with the bolt forward on an empty chamber and the weapon on *SAFE*. The only exception to this is the bow gunner, who will charge his weapon in the boat when directed to fire. He must put the weapon on *SAFE* before debarking, and the squad leader must verify this by touch. The soldier can immediately engage the enemy, upon landing, by simply taking the weapon off *SAFE* and charging the chamber.

WARNING
The soldier must NOT take the weapon off *SAFE* and charge the chamber before leaving the boat.

A safety boat is always used during an assault crossing. One safety boat is used for every company flotilla. It contains at least one lifeguard-qualified swimmer (two, if possible) to assist soldiers that may fall into the water. This lifeguard will not wear boots or LBE. The safety boat will also contain a boat hook and a float with an attached line for rescuing troops in the water. Rocket-propelled lifelines will be included, if available. At a minimum, the boat commander is equipped with night-vision goggles. The crew of the safety boat comes from the supporting engineer force that provides the boats and boat crews and consists of eight paddlers, the coxswain/commander, a medic, and one or more lifeguards. The crew should also have a radio tuned to the company's frequency.

The safety boat crosses parallel with a flotilla and about 40 meters downstream. Its crew pays out a climbing rope fastened to the near shore as a safety rope and attaches life jackets as floats every four boat lengths (see *Figure 8-10*). When the crew reaches the enemy shore, it ties off the safety rope and then moves back to the center of the river. If a man goes in the water or a boat capsizes, the affected boat group makes a quick radio call on the company's frequency, indicating the number in the water and the boat group calling. The alerted safety boat holds water while its crew looks for troops who are in the water or who are caught by the safety rope.

If a soldier goes in the water, he should immediately remove his helmet and release it. He should then roll onto his back. If he is wearing a Class 5 life jacket, he retains his rifle and LBE. If he is wearing a lesser-quality life jacket, he releases his rifle and LBE and drops them. He then allows the current to carry or float him to the friendly shore. He stays alert for the safety rope and safety boat. If he reaches the safety rope, he wraps his arms in it or clips a snap link to it on his LBE (if he is wearing LBE). He either waits for the safety boat or moves along the rope to the nearest shore.

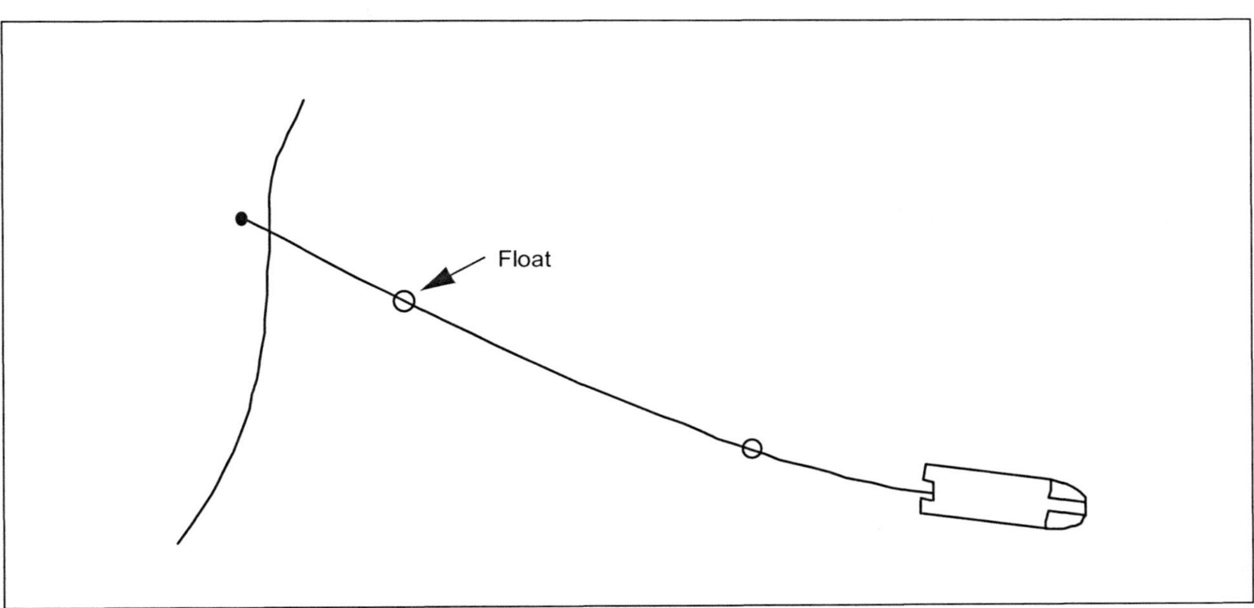

Figure 8-10. Safety boat and guard rope

CHAPTER 9
Engineer Operations

GENERAL

A river crossing requires specific procedures for success because the water obstacle inhibits ground maneuver in the usual way. It demands detailed planning and different technical support than other tactical operations. Extensive use of corps engineer assets are required. It is critical for supporting corps engineers to be totally involved in all facets of the river-crossing operation from initial planning through preparation and execution.

Traffic control is the most vital component of a river-crossing operation. ERPs are used to control traffic flow and permit movement across the river to be unhindered. Engineer units provide positive control and the necessary equipment to ensure maneuver forces successfully and safely cross the river in a timely and synchronized manner.

Contingency operations may require that assault river-crossing assets be used for longer periods of time because fixed bridging is not feasible or readily available. Engineer units must implement techniques that allow the long-term use of assault bridging assets without heavily affecting operations or damaging equipment. These topics will be discussed in detail in this chapter.

ERP OPERATIONS

ERPs ensure the effective use of the crossing means. ERPs and TCPs may be colocated to provide control for the river crossing. The CSC uses them to rapidly organize and move the unit through the crossing area.

The CSC establishes ERPs at the call-forward area and, if enough engineer assets are available, at the staging area and the far-shore holding area. He uses additional ERPs only when specific site conditions make it necessary for crossing-area control. ERP personnel need sufficient space to mark an area the size of a raft (mock-up raft), brief crossing procedures, and conduct necessary inspections and rehearsals. A hardstand, such as a rest stop or parking lot, is ideal for this purpose but lacks the overhead concealment usually desired. Some ERP functions may be done at separate ERPs to ensure a smooth and rapid flow of vehicles to the river. In this case, it is essential to maintain communications between ERPs.

Typically, an engineer squad mans an ERP. This maintains unit integrity and provides sufficient personnel and equipment for continuous operations. The crossing-site headquarters establishes direct communication with ERP personnel to control raft-load or individual vehicle movement. Depending on the location and purpose of the ERP, it can be used for the following functions:

- Briefing crossing-unit personnel on procedures, including safety.

- Demonstrating ground-guide signals.

- Inspecting equipment to ensure that it meets the load-class capability of the crossing means.

- Organizing vehicles into raft loads.

- Conducting rehearsals.
- Controlling vehicle movement.

RAFTING OPERATIONS

ERP personnel configure vehicles into raft loads and send them to the river to coincide with the arrival of an empty raft. Engineers brief crossing units before their arrival in the call-forward area to make this happen as rapidly as possible. The briefing covers the—

- Route and its markings through the crossing site.
- Road speed and interval.
- Loading and unloading of rafts.
- Location of passengers while rafting across the river.
- Configuration of the vehicles for the crossing.
- Actions to take for disabled vehicles and the location of the maintenance collection point.
- Hand and arm signals and signaling devices.
- Arm bands or other identification of guides and traffic controllers.
- Issuing, wearing, and returning of life jackets.
- Location of holding areas and alternate routes.
- Location of the casualty collection point.
- Actions to take in the call-forward area.
- Actions to take in case of enemy fire.
- Regrouping of the company in the far-shore holding area.

An engineer from the squad running the ERP can brief vehicle crews and rehearse the movement signals with them. The staging area is an ideal place to do this, minimizing the time and effort spent organizing a crossing unit in the call-forward area. Otherwise, a separate ERP should handle this task.

Figure 9-1 is an example of an ERP at the call-forward area. The engineer squad leader positions himself where he can best control vehicle movement from the call-forward area to the river line. He establishes communication with the crossing-site headquarters. As a crossing unit arrives, the assistant squad leader contacts the unit's commander, who determines the order in which his vehicles will cross. The assistant squad leader then configures individual vehicles into raft loads, while ensuring that the vehicles do not exceed either the weight limit or the maximum dimensions of the raft. He has a space marked out in the exact dimensions of a raft (mock-up raft) for this purpose. An engineer squad member guides the vehicles onto this mockup raft, using the same procedure to be used at the raft's embarking point on the river. At the same time, another engineer inspects the vehicles for the proper load classification and dimensional clearances and chalks the raft-load number on the vehicles. Once cleared through the mock-up, an engineer squad leader releases individual raft loads to the river as directed by the crossing-site headquarters.

Items useful for running an ERP could include—

- A TA1 field phone and an RL39 with wire.
- Two rolls of engineer tape and six stakes.
- Ten traffic markers.
- Flashlights with colored filters.
- Chemical lights.
- Signal flags.

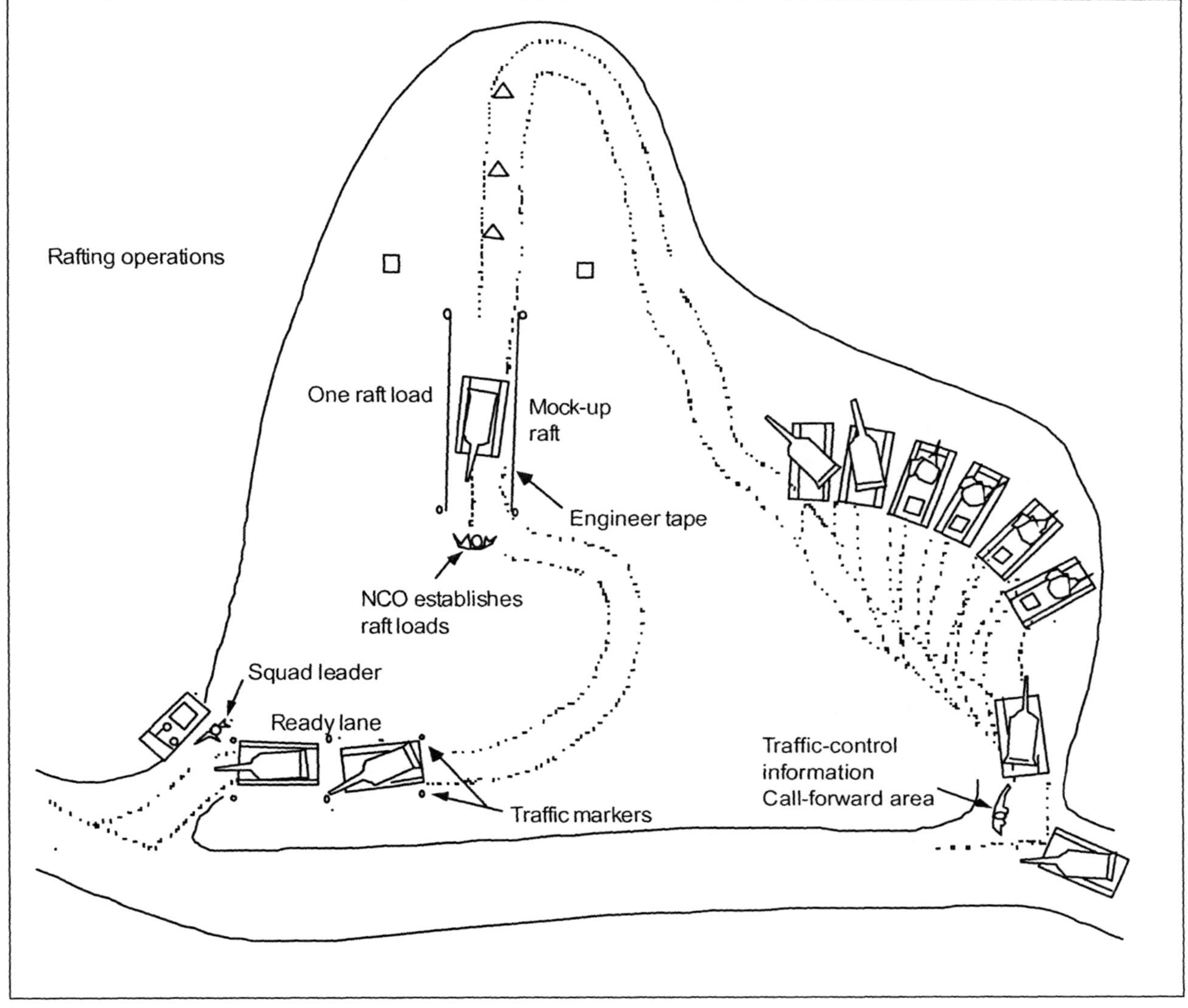

Figure 9-1. ERP layout

- Chalk.

- Camouflage nets and poles.

- Night-vision goggles.

- Sandbags.

BRIDGE OPERATIONS

A bridge operation requires a continuous traffic flow to the river. Units must be briefed and sent to the crossing site quickly.

To accomplish this, engineers brief at staging areas and check vehicle load classification and dimensional clearances. The briefings include the following rules:

- Vehicles will maintain a maximum speed of 9 kph while crossing the bridge.

- Vehicles must not stop on the bridge.

- Operators must not shift or make abrupt changes in speed on the bridge.

- Vehicles will maintain the interval indicated by signs on the side of the road.

- Operators will follow the signals of engineers at ramps and intervals along the bridge.

ERPs may be established along the routes to the crossing site to regulate traffic. A mockup bridge is not necessary at the ERP.

SWIMMING OPERATIONS

For swimming operations, ERP personnel have the necessary briefings and vehicle inspections. Crossing units are responsible for most preparations, but ERP personnel can assist with operations at the predip site that is established nearby and provide recovery assets. A briefing on swimming operations should include—

- The layout of entrance and exit markers.

- Swamping drills.

- Rescue procedures.

- The actions to take in case of enemy fire.

ENGINEER CONTINGENCY BRIDGING OPERATIONS

Organizing and training for war fighting remains the primary mission of Army engineers. However, engineers can be called on to conduct contingency operations. For example, US Army operations in Bosnia included the mission to bridge the Sava River near Zupanja, Croatia, in December 1995. This mission was the largest river-crossing operation since World War II and was conducted under extreme conditions. Seasonal weather caused the Sava river to swell from its normal width of 300 meters to more than 600 meters. Despite harsh conditions, engineers used Chinook helicopters to deploy ribbon-bridge sections into the river while other engineers rebuilt the approaches and successfully bridged the Sava river to allow elements of the 1st Armored Division to cross. As the floodwater receded, engineers built a causeway across the floodplain. As operations in the area continued, the ribbon bridge remained the only crossing means for both military and civilian traffic while preparations were made for fixed bridging.

Versatile engineers provide unique personnel and equipment capabilities that can effectively support complex and sensitive situations in any contingency operation. Therefore, engineer force-projection planning should include the possibility that forces committed to contingency operations may become involved with combat operations. The engineer commander tailors engineer support based on contingency-operation requirements, which may be radically different than supporting combat operations. In many cases, the only difference between a wartime and an engineer contingency operation is the threat level.

Contingency operations may require the same or a greater level of logistics support to engineers as wartime operations. Combatant commanders tailor logistics support to engineers based on theater needs. Logistics efforts are integrated with host-nation or local resources and activities. Engineers invariably get involved with a wide variety of gap-crossing operations that may need flexible logistics support. Critical engineer logistics considerations during contingency operations include—

- The availability of construction equipment.

- A direct-support maintenance capability.

- Repair-parts supply.

- Class IV construction materials.

ASSAULT BRIDGES, LONG-TERM USE

Ribbon-bridge operations normally last no longer than 72 hours. Having the ribbon bridge remain in operation beyond that time frame presents problems for the engineers that normally would not be experienced during a short duration. Equipment maintenance, anchorage systems, constant changes in the water level, and repair of approaches must be considered for long-term use of assault bridges.

MAINTENANCE

As equipment remains in use during crossing operations, maintenance services become more difficult to manage. Time must be made to allow boats and bays to be recovered from the water and completely serviced and checked for unusual wear. The techniques discussed in *Chapter 7* are applicable but must include complete recovery of the equipment and movement back to the EEP where the services can be done. To accomplish maintenance services without jeopardizing bridging operations, boats and their replacements must be carefully managed. This may require procuring more boats than authorized by the table(s) of organization and equipment (TOE) to permit continued crossing operations without distribution for maintenance.

To check and service interior and end bays of the ribbon bridge, it must be broken apart and replacement interior and end bays emplaced. Time for such actions should be incorporated into the bridge-crossing time line and maneuver units notified when the crossing site will be shutdown temporarily. Synchronization of alternating times for crossing sites to be closed for maintenance can proactively re-route traffic flow and prevent major disturbances in movement across the river. To expedite the time required to replace bays needing maintenance and quickly allow traffic to resume crossing operations, engineers prepare replacement bays and boats and stage them before closing the crossing site. Daily checks of the bridge throughout the operation, considerations of the current's velocity and the amount of debris that may affect the bridge's operation, and maintaining vehicle speed across the bridge are critical to prevent damage to the bay's lower-lock devices and roadway-to-bow portion latches.

ANCHORAGE

All military bridges must be held in position by some anchorage system. Short-term anchorage is normally used for assault bridges, but if the bridge is required to remain operational for a longer period, the anchorage must be upgraded to provide long-term support.

The design of any anchorage system is influenced by several factors, including the—

- Width of the river.

- Current's velocity.

- Depth and bottom conditions of the river.

- Height and slope of riverbanks.

- Conditions of the soil.

- Depth of the groundwater table.

- Availability of equipment.

Anchorage of the ribbon bridge must occur if the bridge is used for long-term operation. During short-term crossings, boats maintain the bridge's stability against the current's velocity and keep the bridge from being damaged. However, as time permits, an anchorage system must be emplaced to provide continuous stability and provide relief for the number of boats required. Initially, the anchorage may consist of a combination of shore guys and boats. This method can still allow the bridge to be broken and permit barge river traffic access. Eventually, a semi-permanent anchorage system, such as an overhead cable system, should be emplaced to keep the bridge secure.

The three basic components of all long-term anchorage systems include approach guys, an upstream anchorage system, and a downstream anchorage system. Approach guys are cables that prevent the bridge from being pushed away from the shore as a result of the impact of vehicles driving onto the ramps of the bridge. The upstream anchorage system holds the bridge in position against the force of the current's velocity. The downstream anchorage system protects the floating bridge against reverse currents, tidal conditions, eddies, and high winds or storms that might temporarily alter or reverse the natural flow of the river. The following types of anchorage systems can be used for stabilizing a bridge:

Kedge Anchors

Kedge anchors lie in the streambed and are secured to the bridge bays with anchor lines. They are designed to sink with the stock lying flat and the fluke positioned to dig into the bottom. On hard bottoms, the kedge anchor is useless.

Shore Guys

Shore guys are cables attached from the bridges to a deadman or similar holdfasts on the shore. Shore guys can be upstream or downstream provided that the maximum anticipated current (or reverse current for downstream systems) does not exceed 0.9 MPS. Shore guys can be used for any length of floating bridge provided that a 45-degree angle be maintained between the shore guy and the bridge centerline.

Combination of Kedge Anchors and Shore Guy

A combination system may be used for upstream or downstream anchorage systems in currents less than or equal to 1.5 MPS. When constructing a combination system, attach kedge anchors to every float and a shore guy to every sixth float.

Overhead Cable

An overhead cable system consists of one or more tower-supported cables spanning the river parallel to the bridge. Each end of the overhead cable is secured to the shore, preferably through the use of a deadman. Bridle lines are used to connect each bay of the bridge to the overhead cable. The cable functions like a cable used in a suspension bridge, except that its final working position is inclined toward the bridge because of the force of the current on the bridge.

TC 5-210 provides the specific criteria for the design of an overhead-cable anchorage system, to include the cable design, tower design and placement, and deadman design.

PROTECTIVE SYSTEMS

Floating bridges, particularly those that will remain in place for long periods of time, must be protected against severe weather conditions and enemy destruction. If flood conditions or heavy debris hamper bridging operations, removing of interior bays will reduce the lateral pressure on the bridge and allow the debris to pass downstream. If losing the bridge is imminent, release an end section and securely anchor the bridge parallel to the shore until conditions permit resuming bridging operations. As the river's width increases, simply add more interior bays to the bridge to compensate.

The enemy may attempt to destroy floating bridges in a variety of ways, including air attacks, land attacks, underwater demolitions teams, floating mines, or assault boats. It is necessary to construct floating protective devices to prevent waterborne forces from damaging or destroying the bridge. The three types of floating protective systems are as follows:

Antimine Boom

This device is designed to stop any mines that are sent downstream toward the bridge. The antimine boom is placed far upstream to protect the other protective devices as well as the bridge. It consists of a number of logs or other large floating structures attached to a cable running across the river. Concertina is normally placed along the length of the boom.

NOTE: Before using timber logs or railroad ties, ensure that they are not waterlogged and will float.

Impact Boom

The impact boom is designed to withstand the impact of large natural or man-made debris and stop the enemy from attacking the bridge by boat. It is constructed by placing a series of floats and cables across the river. The cables absorb the impact of the debris or boat and restrain it until it can be removed or destroyed.

Antiswimmer Net

This net is used to stop swimmers or underwater demolition teams from reaching the bridge. The net can be constructed by suspending a mesh or net barrier from an anchorage cable to the river's bottom. Concertina may also be connected to the cable and net to prevent swimmers from climbing over the net. The net must be firmly affixed to the river's bottom or enemy divers can easily go under the net. The antiswimmer net should also be placed on the downstream side of the bridge to prevent enemy divers from reaching the bridge from downstream.

Army diver teams can assist in emplacing the protective devices and test them to ensure they are able to prevent penetrating the bridge.

APPROACHES

Over a period of time, traffic flow at the same location will eventually wear the approaches down and make them unusable. Engineers incorporate repair of the entry and exit banks and the approaches leading to the crossing site into the crossing-operation plan. Initially, the approaches may be suitable to receive heavy traffic with little effect, but implementation of reinforcing the approaches must be done for long-term traffic. When inspecting approaches, consider the following:

- The steepness of the approaches.

- The ruts or gullies along the approaches, particularly in a floodplain area.

- Water-level conditions and expected changes due to weather or seasonal conditions.

- The location of alternate approaches (alternate crossing sites) to allow for the repair of existing approaches.

Matting and rock or gravel are the best suitable materials to use to support the approaches. Maneuver units that will have to conduct long-term crossing operations should develop procedures to requisition and deliver these materials to identified crossing-site locations. Reconnaissance teams can locate local quarries where rock and gravel can be obtained through coordination with the host country.

New techniques for constructing bridge approach roads include using fabric as a reinforcement across soft soil. An impervious, neoprene-coated, nylon-woven membrane can be placed between a stone aggregate and the soft-surface soil to allow the ground to withstand heavy traffic. The most important feature that a reinforcing fabric membrane can offer is improving soil stability and strength, which creates smaller deformations from vehicle traffic than soils acting alone.

LONG-TERM GAP-CROSSING C^2

More than any other mobility task, gap crossing involves managing combat power, space, time, and terrain. The controlling headquarters must be flexible enough to react to any changes in the tactical situation and scheme of crossing. This is particularly difficult when involved with long-term operations in the same area of operations. Management of the crossing site, enemy considerations, traffic-control measures, and CSS must be synchronized for long-term activities and must not be based on less than a 72-hour period.

Management

Traffic and movement control remain the responsibility of the C^2 headquarters. Activities may direct that another unit take over the crossing operation and equipment as a whole or bring their own crossing equipment and personnel to relieve the existing units and permit them to move forward. All aspects of the operation must be covered when handing over the crossing site to the gaining unit—just as though they were conducting the crossing for the first time.

Enemy Considerations

Operation of a single crossing site over an extended period of time increases the possibility of enemy interdiction. The possible use of nuclear or chemical weapons against friendly crossing activities impacts on control procedures. To prevent the friendly elements from becoming targets, forces must cross the gap as swiftly as possible to minimize the concentration of forces on either side of the gap. The controlling headquarters also varies the crossing-site location to reduce enemy threat interdiction.

Traffic-Control Measures

Staging- and holding-area control must be maintained. These areas must be located far enough away from the gap to facilitate rerouting and the use of alternate roads to crossing sites. Staging and holding areas on the far shore must be developed to handle the traffic as vehicles travel back across. New routes may be constructed and existing routes upgraded to improve traffic flow. Staging areas must be able to provide the following:

- Cover and concealment.

- A sufficient area for vehicle and equipment dispersion.

- Easy accessibility.

- Sufficient trafficability to prevent delays caused by increased traffic flow within the area of operation.

Combat Service Support

In a normal gap-crossing situation, the committed combat forces will be temporarily separated from their full CSS. For long-term gap-crossing operations, increased traffic flow for the service-support vehicles must be considered and controlled. Sufficient crossing sites and designated crossing times can ensure that priority is given to field trains and ensure that timely resupply operations are not hindered. Recovery of nonmission-capable equipment presents an additional problem for recovery teams transporting the equipment back to the near shore for direct-support maintenance support. Additionally, recovery resources should continue to be provided at both sides of the crossing sites so they can quickly recover a vehicle unable to cross and prevent delays.

MULTIROLE BRIDGE COMPANY (MRBC)

Today's Army must be able to respond to an increasing array of potential employments in war or in contingency operations. With the new integrated information-processing systems, river-crossing opportunities will be quickly identified and execution will rely heavily on the rapid-response capability of bridge companies.

Current bridge companies are tailored around their specific floating or fixed-span bridges, without regard for their inherent conceptual, operational, and organizational commonalty. Both types of companies are similarly organized with a company headquarters, two bridge platoons, and a support platoon. The fixed-bridge company cannot transport ribbon-bridge sections on its 5-ton dump trucks, nor does it have the required bridge boats. The float-bridge company, on the other hand, can readily transport fixed-bridge components on its transporters, but only if the loads are palletized.

Whenever both types of bridges are needed, both types of companies must deploy. Since a company owns it own unique bridge, sufficient companies must be deployed to cover all possible gaps. When a bridge company has expended its bridging materials, it reverts to a transportation mission.

The MRBC is a modular company with both dry- and wet-gap capabilities that will employ both types of bridges as needed, based on METT-T. Each bridge platoon can transport and employ both types, as ordered. The MRBC is a versatile, flexible, and modular unit that can meet the demands of the maneuver commanders. The developmental HDSB is also fully compatible with the MRBC concept. The HDSB will replace the MGB as the primary fixed bridge employed by this company, relegating the MGB to a supplemental role.

ORGANIZATION

The MRBC will be a combination of a MGB company and an AFB company. The MRBC's structure consists of a company headquarters, two bridge platoons (one MGB and one AFB), and a support platoon. The support platoon consists of a platoon headquarters, an equipment maintenance section, and a bridge-site section. The new bridge transporter, the improved common bridge transporter (ICBT), is specifically designed to function in the MRBC. The palletized-load-system (PLS) trailer will be procured as a part of the HDSB to allow the transporting of both bridges simultaneously.

BASIC CONCEPT

A typical operational mission would begin with a platoon responding to a mission with an initial basic load of the desired bridge. When it completes the bridge, the platoon moves to the next site, either picking up the next required bridge along the way or finding the bridge cached at either the engineer bridge park or the site it will be emplaced. As the forces advance, the bridges become the responsibility of the engineer units in the communications zone. As prefabricated bridging is replaced by nonstandard or more permanent bridging, the platoon responds to bridge-retrieval missions. Retrieved bridges will reenter the supply system or be stored in the unit's bridge supply yard. Bridging sets are a supply commodity and are handled as any other Class VII supplies (major end item).

The MRBC's capabilities will give commanders the ability to quickly maneuver and respond, through either fixed or float bridging or a combination of bridging appropriate for the mission. This newly structured MRBC operates within a "lane" on the battlefield (an area defined by a maneuver brigade's area of operation) and provides the necessary bridging for multiple axes of advance. Trailing closely behind combat forces, the MRBC moves forward when necessary to support assault crossings. Additional MRBCs, provided from corps assets, may be assigned to the lane when gap-crossing requirements exceed the capabilities of a single MRBC.

IMPLICATIONS

Because bridge sets are exceptionally heavy and tall and have many parts (some are small and easy to lose), they should be placed on pallets and shrink wrapped for transportability and accountability. The procedure to requisition and deliver bridges is essentially unchanged. The number of bridges needed is unaffected, since it is based on METT-T and not the bridge-company organization.

Bridges are loaded on bridge adapter pallets and boats that have the improved boat cradle, which allows the loads to be placed on the ground without damage. The pallets and cradles remain at the EEP until the bridge or boat is recovered. Treating bridging as a commodity instead of a bridge company TOE property emphasizes the critical maneuver planning that is involved by the maneuver staff and not exclusively the engineer staff. MRBC engineers cannot exist on the battlefield without support from the following branches:

- Quartermaster—to order and track bridge stocks.

- Transportation—to haul replenishment bridge sets forward to engineer bridge parks.

- Ordnance—to repair transporters, erection boats, and bridge components.

TRAINING

Engineer units' training must reflect the dual bridge capability. Currently, bridge crewmen receive advanced individual training on both fixed and float bridges. However, once they are assigned to a unit, their collective training is only on the single-type bridge of that company. Under the MRBC concept, individual bridge crewmen and leaders must maintain their proficiency on both types of bridges. Bridge specialists must continue to be proficient in all types of prefabricated bridges. Critical branch interaction during war-gaming exercises must consider employing the MRBC in current training by implementing a variety of missions, either sequentially or simultaneously, to become more accustomed to its employment. Additionally, engineer company-grade officers and noncommissioned officers (NCOs) will have increased responsibility and will need to improve their technical proficiency.

Final decisions as to the size and end-state composition of the MRBC are not being implemented yet; however, many units can expect to have the HDSB and the improved ribbon bridge carried on HEMTT-chassis transporters. Interim organizations will include basic loads of the medium-girder, Bailey, and ribbon bridges, which are carried on their current transporters. Whatever the organization's development stage, the MRBC will not be dedicated to only one bridge type.

APPENDIX B
Engineer-Planning Calculations

GENERAL

This appendix addresses the detailed engineer planning necessary for a river-crossing operation. The charts and overlays that are used to synchronize and control execution of the crossing are shown in *Figures B-1 through B-8 and Tables B-1 through B-4, pages B-4 through B-13.* The H, as used in the figures in this chapter, stands for H-hour; this is the specific hour the assault phase begins (see *FM 101-5-1*).

ENGINEER PLANNING

Initial engineer planning at corps and division levels focuses on providing sufficient engineer assets to handle crossing requirements. The terrain teams at division and corps levels maintain the terrain database that provides potential crossing sites and river widths. The division engineer uses this information to construct a site overlay (see *Figure B-1, page B-4*). He labels assault and rafting or bridging sites and shows the site capacity and the estimated preparation time for each site (from the terrain database).

Preparation time is the time required to improve routes and riverbanks to support the units that will use the site. It also includes the time required to construct rafts and bridges. Rafting-site capacity is the number of raft round trips per hour. The engineer calculates rafting-site capacity by multiplying the number of raft trips per hour by the number of rafts and the number of centerlines at the site (see *Table B-1, page B-5*). Centerlines must be at least 100 meters apart. Each assault company needs 200 meters of river frontage. *Figure B-1* shows the determination of rafts per hour and the capacity of the assault site for the division crossing overlay. The site overlay provides additional details necessary to ensure that each brigade has sufficient potential crossing sites within its boundaries. *Table B-2, page B-5*, provides planning factors for assault-boat operations.

Rules of thumb for making this determination follow:

- A brigade requires 31 assault boats to cross a battalion with three companies in the first wave. With 70 boats, it can cross two battalions at once. Generally, the boats with the corps bridge companies can handle these requirements.

- A brigade requires two bridges or the equivalent bridging configured into rafts.

The engineer planner uses the above rules of thumb to task-organize engineers that are supporting each crossing area. The division engineer then develops a rough crossing time line using pure battalions. This provides sufficient information for division planning, without requiring detailed knowledge of the brigade's plan. *Table B-3, page B-5*, provides necessary planning factors (field trains are not included). *Figure B-2, page B-6*, illustrates a crossing time line using 6-float rafts.

The brigade headquarters does the majority of planning for the detailed crossing. During the mission analysis, the brigade engineer also develops a crossing time line to provide initial buildup-rate information to the maneuver planners when they outline possible schemes of maneuver. This time line is the same as the time line that is developed at the division and may be provided by the division engineer.

Once the commander identifies the COAs to develop, the staff engineer develops crossing-area overlays for each (see *Figure B-3, page B-7*). These overlays are developed by using information from the site overlay, along with additional terrain data. The crossing-area overlays show staging areas, holding areas, call-forward areas, and routes for each crossing site included in the COA. A crossing-area overlay is necessary for each COA. The overlay for the COA eventually selected is later modified by adding ERPs, TCPs, and crossing-area-headquarters information and is used to support the operation.

When maneuver planners develop COAs, they assign crossing sites and the order of crossing to units and task-organize the pure maneuver battalions into TFs. The engineer uses this information to construct a crossing time line for each COA. He calculates the number of vehicles and raft loads for each unit using pure company figures from *Table B-4, page B-8*. The company's raft requirements do not include the field trains. The engineer then calculates the crossing time for the unit by using the crossing capacity of the site assigned to it. The crossing time line shows these crossing periods, by site, based on the order of crossing. The engineer then develops a detailed crossing time line based on the task organization (see *Figure B-4, page B-9*).

During the comparison of the COAs, the engineer uses time lines, brigade-site overlays, and crossing-area overlays to demonstrate the differences in the crossing plans. After the commander has selected the COA for the mission, the staff converts

it into a detailed plan. The engineer develops a vehicle-crossing-capability chart.

He starts by displaying the capabilities of each crossing site in terms of raft loads per hour (rafting operations) or vehicles per hour (bridging operations). Since the crossing rate for rafts is less during darkness, each site shows total raft trips separately (during day and during night). An example of the product of this first step is shown in *Figure B-5, page B-10*).

The engineer determines the crossing requirements using the factors from *Table B-4*. He then blocks out the crossing periods for all units based on site assignment, site capability, and the crossing order in the scheme of maneuver. After adding the units' crossing periods to the chart, he coordinates the plan with the S3 to ensure that the units will arrive on the far shore by the times they are needed (see *Figure B-6, page B-11*). If not, the S3 and engineer work together to adjust the crossing order of subordinate units. The basic technical information remains constant as different crossing sequences are checked until one meets far-shore requirements. The vehicle-crossing-capability chart is the primary tool for finalizing the crossing plan.

After the crossing order has been established, the engineer develops the crossing-synchronization matrix (see *Figure B-7, page B-12*). This is the tool that the CAC and CAE will use to synchronize the execution of the crossing. It is constructed as a chart, with the unit's locations and activities displayed by time on the upper half and terrain occupation displayed by time on the lower half. The staff can follow each unit's location as the operation progresses and can easily see potential conflicts resulting from changes. The matrix also provides critical information for traffic control.

The crossing synchronization matrix is constructed backwards by first portraying the units' crossing times established from the vehicle-crossing-capability chart, then by using road movement times to show route usage and staging-area times. The time required for the crossing of the assault force is also included. Once all of the units are displayed, the same information is transferred to the lower terrain portion of the matrix.

The staff immediately resolves any conflicts they discover while preparing the matrix.

The final engineer planning step is developing the engineer execution matrix (see *Figure B-8, page B-13*). It displays subordinate units' task assignments by time. It is useful both for tracking unit execution and for aiding decisions if changes to the plan are required.

Site	River Width in Meters	River Front-age	Number of Rafts/ Center-line	Number of Center-lines	Raft Round Trip/ Hour/ Center-line	Total Raft Round Trip/ Hour	Prep Time/ Total Raft/ Hour
1	175	300	2	3	5.4	32	1-hour prep 32 rafts/hour
Assault A and 2	180	410	2	4	5.4	43	3-hour prep 43 rafts/hour
3	160	400	2	4	6	48	0.5-hour prep 48 rafts/hour
Assault B and 4	140	600	1	6	6.7	40	1-hour prep 40 rafts/hour
5	150	250	2	2	6	24	1.5-hour prep 24 rafts/hour
6	150	310	2	3	6	36	2-hour prep 36 rafts/hour
7	165	300	2	3	6	36	0.5-hour prep 36 rafts/hour
Assault C	150	400	NA	NA	NA	NA	2 co
8	150	350	2	3	6	36	2-hour prep 36 rafts/hour
Assault D and 9	160	610	2	6	6	72	0-hour prep 72 rafts/hour
10	170	375	2	3	6	36	0-hour prep 36 rafts/hour
Assault E	175	650	NA	NA	NA	NA	
11	180	300	2	3	5.4	32	1-hour prep 32 rafts/hour
Assault F	175	685	NA	NA	NA	NA	
12	170	250	2	2	6	24	1-hour prep 24 rafts/hour

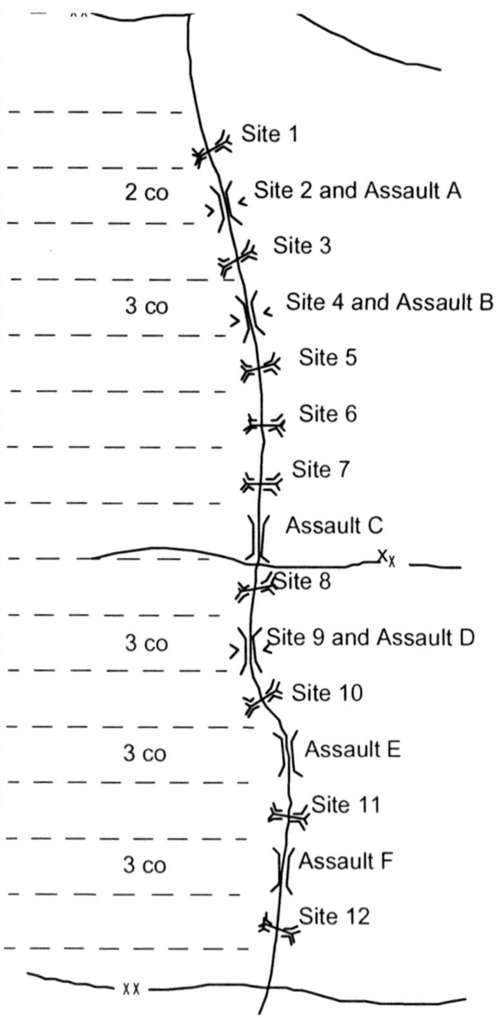

NOTES:

1. Obtain the river's width from the terrain database and/or reconnaissance.

2. See *Table B-1* for the number of raft round trips per hour per centerline and the number of rafts per centerline.

3. Determine the amount of available river frontage from the terrain database and/or rec

Figure B-1. Division-site overlay

Table B-1. Raft-centerline data

River Width in Meters	Round Trip in Minutes	Number of Raft Trips per Hour	Number of Rafts
75	7	8.6	1
100	8	7.5	1
125	9	6.7	1
150	10	6.0	2
175	11	5.4	2
225	12	5.0	2
300	16	3.75	3 to 5
NOTE: Planning times are for current velocities up to 1.5 MPS			

Table B-2. Boat-planning factors

Equipment	Characteristic	River Width		
		75 Meters	150 Meters	300 Meters
Pneumatic assault boat with an OBM	Minutes per round trip Trips per hour	3 20	4 15	5 12
Pneumatic assault boat without an OBM	Minutes per round trip Trips per hour	4 15	6 10	10 6

NOTES:
1. **Factors are averaged based on load/unload time and safety.**
2. **Planning times are for current velocities up to 1.5 MPS. For faster current velocities, classification must be reduced to a caution or risk crossing, and an engineer analysis must be made of the actual site conditions before planning times may be assessed.**

Table B-3. Unit rafting requirements

Units	Vehicles	Raft Trips Required		
		4 Bays	5 Bays	6 Bays
Armored battalion	161	119	101	86
Mechanized battalion	153	112	65	55
FA battalion	165	97	61	52
Engineer battalion (ERI)	139	77	59	50
ACR	208	171	110	98

NOTE: Assume that current velocities are less than 0.9 MPS and that battalions/regiments are at 100 percent MTOE strength.

	H	H+1	H+2	H+3	H+4	H+5	H+6	H+7	H+8
1st brigade Site 1 32 raft/hour		Prep	Armd bn (86 rafts)			FA bn (32 rafts)			
Site 3 48 raft/hour		Prep	Mech bn (55 rafts)	Engr bn (50 rafts)	Armd bn				
2nd brigade Site 8 36 rafts/hour		Prep			Mech bn	FA bn			
Site 9 72 rafts/hour		Armd bn	Mech bn	Engr bn					

Figure B-2. Rough division-crossing time line

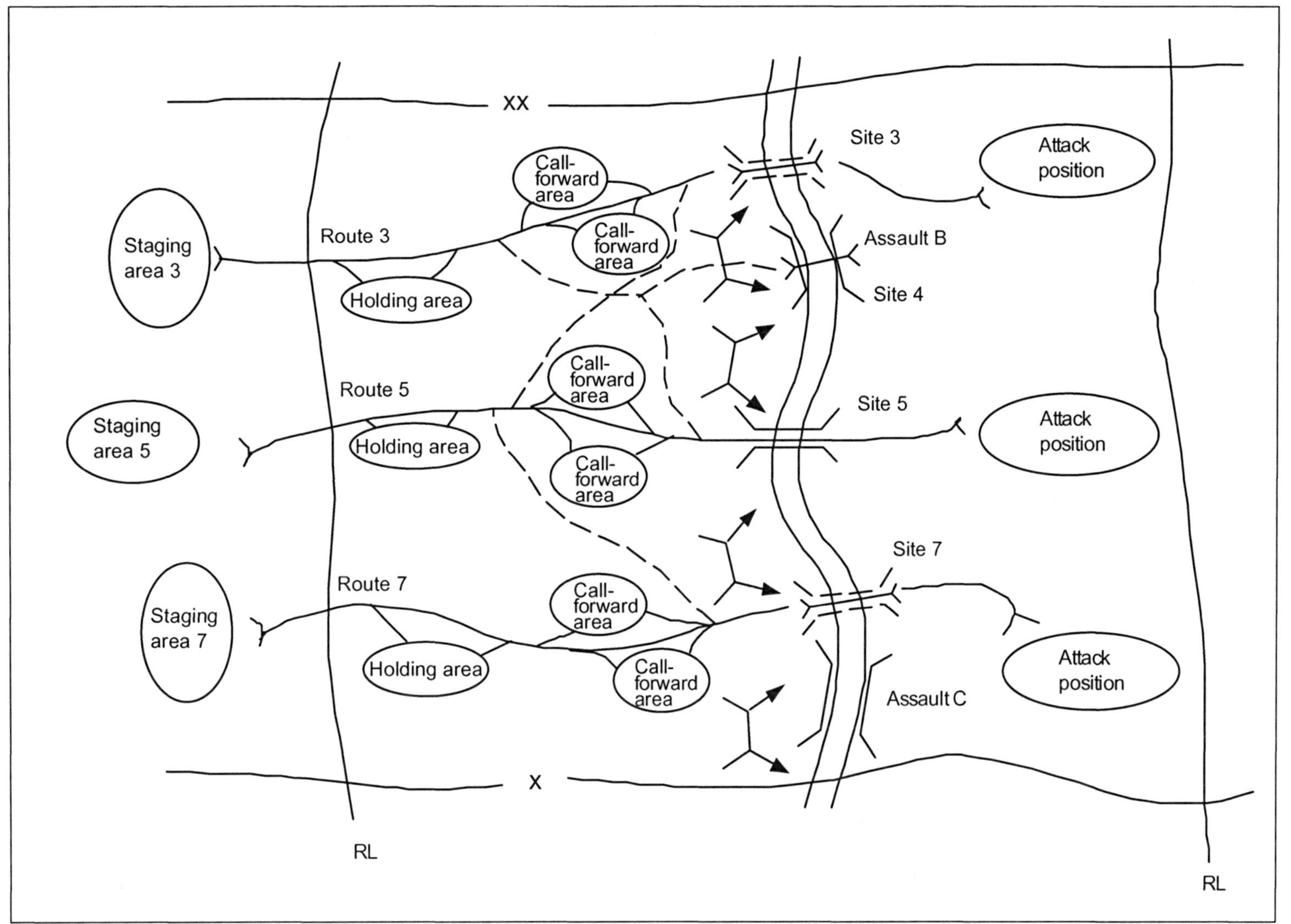

Figure B-3. COA crossing-area overlay

Table B-4. Pure company rafting requirements

Units	Vehicles	Rafts Required		
		4 bays	5 bays	6 bays
Tank company	15	15	14	14
Mechanized company (Bradley)	15	14	7	7
Armored TF HQ	6	4	4	3
Mechanized TF HQ	6	4	4	3
Mortar platoon	8	3	2	2
Scout platoon	6	3	2	2
Engineer platoon	5	3	2	2
Division cavalry troop	24	23	16	16
ACR troop	27	25	18	17
ACR squadron HQ	6	4	3	2
155-SP artillery battery (division)	18	16	9	9
ACR tank co	15	15	14	14
FA battery (ACR)	13	13	10	7

	H	H + 1	H + 2	H + 3	H + 4	H + 5	H + 6	H + 7	H + 8	H + 9	H + 10
Site 1	Not used										
Site 2	Not used										
Site 3		Prep	Mech TF-1		Engr bn						
Assault B	TF-1										
Site 4		Prep		Alternate site							
Site 5		Prep	Armd TF-1	Armd TF-2	FA						
Site 6	Not used										
Site 7		Prep	Mech TF-2	ADA							
Assault C	TF-2										

Figure B-4. Brigade-crossing time line for a COA

Site	Crossing Means	Trips/Hour		Beginning Morning Nautical Twilight							
		Day	Night	H	H + 1	H + 2	H + 3	H + 4	H + 5	H + 6	
3	8 six-bay ribbon rafts; convert to bridge	40	26		Site prep & const	26 rafts	40 rafts	Const bridge	Bridge		
5	Bridge	200	200 Vehicles/ hour			Site prep/ const bridge	200 vehicles	Bridge 200 vehicles/hour			
7	6 six-bay ribbon rafts; convert to bridge	36	24		Site prep & const	24 rafts	48 rafts	Const bridge	Bridge		

Figure B-5. Initial vehicle-crossing capability

Site	Crossing Means	Trips/Hour		Beginning Morning Nautical Twilight						
		Day	Night	H	H + 1	H + 2	H + 3	H + 4	H + 5	H + 6
3	8 six-bay ribbon rafts; convert to bridge	40	26			Mech TF-1	Mech TF-1 / Engr bn HQ		Follow-on forces	Follow-on forces
					Site prep & const	26 rafts	40 rafts	Const bridge	Bridge	Bridge
5	Bridge	200	200			Armd TF-1	FA bn / Armd TF-2	Field trains	Field trains	Field trains
		Vehicles/hour			Site prep/const bridge	85 veh	200 vehicles	Bridge 200 vehicles/hour	Bridge 200 vehicles/hour	Bridge 200 vehicles/hour
7	6 six-bay ribbon rafts; convert to bridge	36	24			Mech TF-2	Mech TF-2	Mech TF-2	Follow-on forces	Follow-on forces
					Site prep & const	24 rafts	48 rafts	Const bridge	Bridge	Bridge

Figure B-6. *Final vehicle-crossing capability*

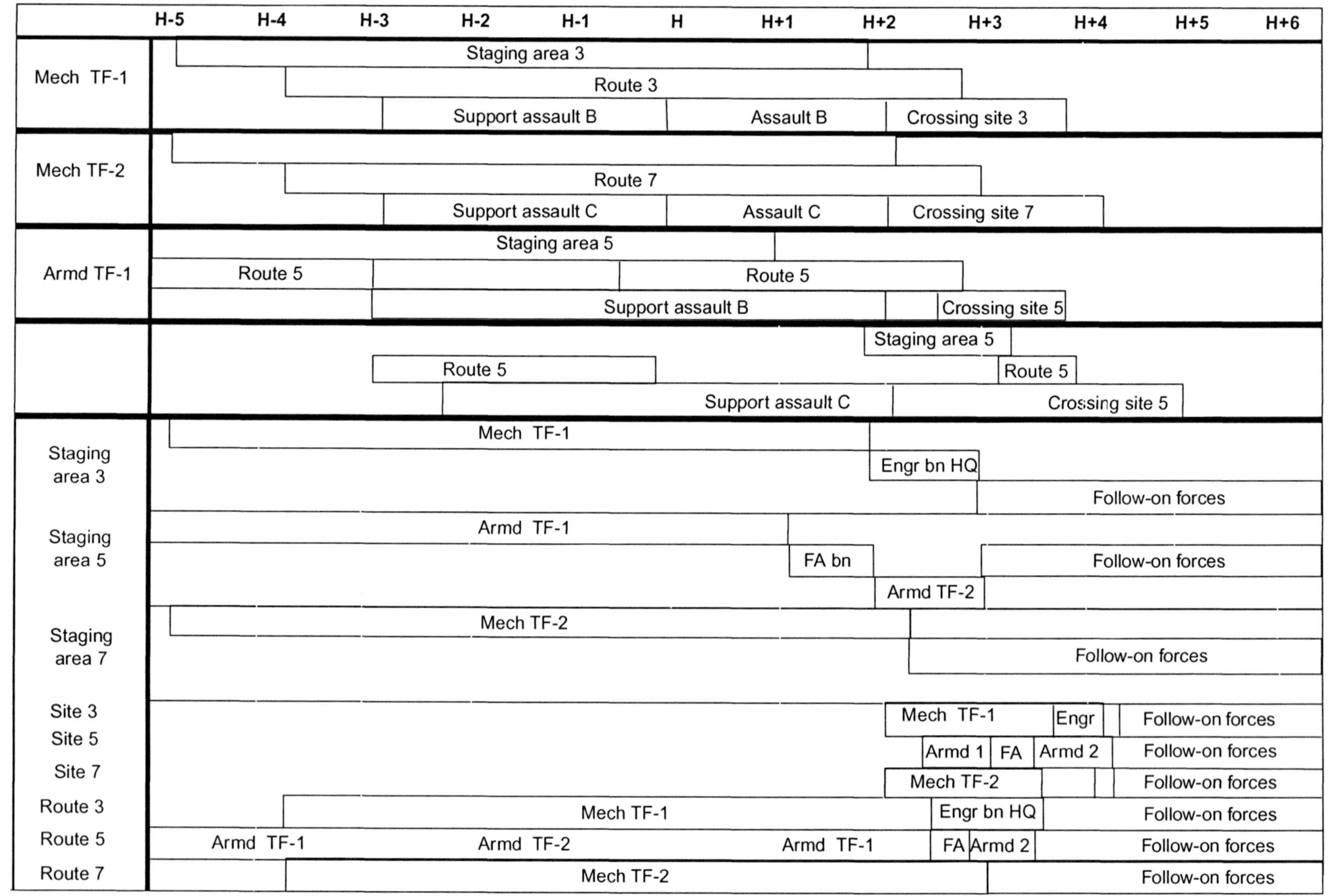

Figure B-7. Crossing-synchronization matrix

	H-3	H-2	H-2	H	H+1	H+2	H+3	H+4	H+5
A/237	Move to site B. / Move to site 3.	Prep RB15s. / Establish ERPs.	Execute assault-boat operations; assault site B.			Prepare site 3.	Perform route maintenance of route 3. Operate crossing site 3.		
B/237	Move to assault site C.	Position and prepare boats.	Execute assault-boat operations assault site C.				Perform route maintenance of route 7.		
C/237		Move to site 7.	Establish ERPs.		Prepare site 7.		Operate crossing site 7.		
D/237		Move to site 5.	Establish ERPs.		Prepare site 5.		Perform route maintenance of route 5. Operate crossing site 5.		
203 AFB co	Deliver assault rafts.		Move to equipment park 3.		Build rafts, site 3.		Operate rafting site 3.		Construct bridge, site 3.
204 AFB co	Deliver assault rafts.		Move to equipment park 5.		Construct bridge, site 5.		Operate bridge site 5.		
205 AFB co	Deliver assault rafts.		Move to equipment park 7.		Build rafts, site 7.		Operate rafting site 7.		Construct bridge, site 7.

Figure B-8. Engineer execution matrix

APPENDIX C
Crossing Means

GENERAL

Crossing means is the equipment used to carry a force across a water obstacle. This equipment is specially designed to operate within certain limits, and commanders must understand these limits if the force is to cross safely.

A safety matter that affects operational use is the load capacity of rafts, bridges, and equipment. The quantities shown in *Table C-1, page C-4*, are the normal capacities or the design capabilities. In exceptional circumstances, certain safety factors or margins allow an increase in the load. These capacities have been deliberately omitted here because they are not intended for use in operational planning. The standard or design capabilities are provided for normal crossings. The exceptional category is intended for special situations using the terms caution or risk crossings.

In addition to the command decision required to employ caution- and risk-crossing loads, commanders must consider the physical status of the equipment. Thus, crossing-area or crossing-force commanders obtain a professional judgment from an engineer. The commander weighs these factors with the tactical needs before directing an increase in the load, keeping in mind that the equipment may be lost for future use. River crossings have three categories:

- Normal crossing—the vehicle's classification number is equal to or less than the bridge's, vehicles maintain 30-meter intervals on fixed or floating bridges, and the vehicle's speed must not exceed 24 kph. Sudden stopping or acceleration is forbidden.

- Caution crossing—vehicles with a classification exceeding the capacity of the bridge by 25 percent are allowed to cross under strict traffic control. The caution classification number of standard fixed or floating bridges may be obtained from *FM 5-34*, *TC 5-210*, or other appropriate technical manuals (TMs). The crossing requires that vehicles remain on the centerline and maintain 50-meter intervals. The crossing requires that vehicles do not exceed 13 kph, stop, accelerate, and shift gears.

- Risk crossing—may be made only on standard, prefabricated fixed and floating bridges and in the greatest emergencies. The vehicle moves on the centerline and is the only vehicle on the bridge. The crossing requires that vehicles do not exceed 5 kph, stop, accelerate, and shift gears. The vehicle's classification number must not exceed the published risk classification for the bridge type being crossed. After the crossing and before other traffic is permitted, the engineer officer reinspects the entire bridge for any damage.

DESCRIPTIONS OF CROSSING MEANS

This appendix supplements a general description of the crossing means discussed in *Chapter 3.* It provides a pictorial review as well as equipment-capability tables useful in selecting crossing means and planning crossing operations.

Available crossing means dictate both crossing operations and the force-buildup rate on the far shore. Before the commander develops his tactics, he must understand how the available crossing means impact his ability to mass forces on the far shore .

The following are the crossing means that the military use to cross a river:

- Fording vehicles.

- Amphibious vehicles.

- Aircraft.

- Boats.

- Assault launched bridges.

- Rafts.

- Bridges.

FORDING VEHICLES

Combat vehicles can ford shallow rivers that have a limited current velocity and stable beds. Some vehicles have kits to increase fording depth. Fording is possible for current velocities that are less than 1.5 MPS. Riverbeds at fording sites must be firm and free of large rocks and other obstructions. Vehicle-operator manuals contain specific depth capabilities and required adaptations. The AVLB and Wolverine can be used to assist fording vehicles in deep water.

AMPHIBIOUS VEHICLES

Some combat vehicles can swim. Bank entry and exit points must be clear of obstructions and have slopes consistent with the vehicle's capabilities. The current's velocity sets limits. Crews of amphibious vehicles prepare and inspect each vehicle before entering the water. Engineer assistance, including recovery vehicles and standing cables, maximizes swimming opportunities.

AIRCRAFT

Army aircraft are the primary crossing means for dismounted infantry. Helicopters also lift other crossing assets from rear areas to the river and carry essential combat support and critical resupply across it. See *Table C-2, page C-5*, for characteristics of external loads for aircraft.

BOATS

Pneumatic assault boats are the alternate crossing means for dismounted infantry and accompanying elements. For light infantry, assault boats may be the only means required if air resupply is available. They carry 12 assault troops and a two-man engineer crew in a silent or powered crossing.

ASSAULT LAUNCHED BRIDGES

The AVLB is an organic engineer asset that travels with maneuvering armored and mechanized infantry formations and can quickly gap up to 15 meters for 70 MLC vehicles. The assault launcher can launch the bridge without exposing bridge personnel to enemy fire and can retrieve the bridge from either end (see *Figure C-1 and Table C-3, page C-6*).

The Wolverine will eventually replace the AVLB. The Wolverine will consist of an M1-series Abrams tank chassis modified to transport, launch, and retrieve a MLC 70 bridge. The bridge will be capable of spanning at least a 24-meter gap (see *Figure C-2 and Table C-4, page C-7*).

RAFTS

Heavy rafts are often the initial crossing means for tanks and other fighting vehicles. They are faster to assemble than bridges and can operate from multiple sites to reduce their vulnerability. The two types of heavy rafts in the Army system are the ribbon and M4T6 (see *Figures C-3 and C-4, page C-8, and Tables C-5 through C-9, pages C-9 through C-12*). The ribbon raft is fielded to engineer units while the M4T6 is maintained in war stocks only.

BRIDGES

Rafts alone cannot handle the total volume of traffic in the needed time. Floating bridges are the primary means to cross the force and its supplies rapidly. The same units that provide heavy rafts also provide float bridges. They often assemble bridges from the rafts used earlier.

The ribbon bridge is the primary assault bridge because it is quick to assemble (see *Figures C-5 and C-6, page C-13, and Tables C-10 and C-11, page C-14*). The M4T6, currently maintained in war stocks only, would replace the ribbon bridge to allow the ribbon bridge to continue to move forward with the advancing force. Because it is man-power intense, the M4T6 is slower to assemble than the ribbon bridge (see *Figure C-7, page C-15, and Tables C-12 and C-13, pages C-15 and C-16*). Preassembly of the M4T6 floats in rear areas significantly reduces the final assembly time on the river.

Fixed bridges rest on the riverbanks and intermediate supports instead of floating on the water. They span ravines as well as rivers. They have limited use for the initial assault because they are slow to assemble and vulnerable to enemy action. Where appropriate, fixed bridges supplement or replace float bridges. Engineers also use fixed bridges to repair existing damaged bridges.

The rapid construction characteristics of the MGB versus the Bailey bridge provide it with a better capability that can be used well forward in the main battle area. Since the Army does not currently have a tactical dry-gap capability longer than 60 feet, using the MGB in this role becomes an important operational consideration (see *Figures C-8 and C-9, page C-17, and Tables C-14 through C-18, pages C-18 through C-21*). The primary role of the MGB is for tactical bridging in the brigade area, while the Bailey bridge is used primarily as a LOC bridge. As the tactical situation permits, the MGB is removed and replaced by Bailey, timber, or steel bridges.

The Army is currently in the process of developing the HDSB to replace the MGB. The HDSB provides tactical bridging for a gap-crossing capability of 40 meters without intermediate supports for wheeled vehicles up to MLC 96 and tracked vehicles up to MLC 70 (see *Figures C-10 and C-11, pages C- 21 and C-22*).

The M2 Bailey bridge is a truss bridge manually assembled by connecting panels end to end. It is used in forward areas to replace assault bridging and the MGB. The Bailey bridge system is highly labor intense but also highly versatile. In some cases, the Bailey bridge is the only tactical bridge suitable for long spans and heavy loads because it can be assembled in multiple heights and widths. The Bailey bridge is maintained in war stocks both in the US and outside continental US (OCONUS). The bridge system can also be assembled as a railway bridge, thus providing a relatively rapid-repair capability (see *Figure C-12, page C-22, and Table 19, page C-23*).

In arctic regions and areas that experience seasonal winter weather, a consideration that cannot be overlooked is "ice bridging". Ice bridging is the use of bridging over a thick layer of ice that covers a wet gap, such as a lake or river (see *Figures C-13 through C-15, pages C-24 and C-25, and Tables C-20 through C-23, pages C-25 through C-27*).

Table C-1. Equipment-characteristic chart

Equipment	Allocation	Transportation	Capabilities	Assembly/ Propulsion	Remarks/ Limitations
Pneumatic, 15-man assault boat	L-series TOE provides— • 18 per ribbon bridge co. • 27 per corps float bridge co. • 9 per sep bde engr co. • 27 per corps ribbon bridge co. • 80 per assault-boat team. • 21 per MGB co. • 21 per M2 co.	A 2 1/2-ton truck can carry 20 deflated boats. An inflated boat can carry 8 men. A deflated boat weighs 132 kg.	The boat can carry 12 soldiers and 3 engrs with paddles or 12 soldiers and 2 engrs with an OBM or 1,531 kg of equipment.	Inflation time is 5 to 10 minutes with pumps. Paddle speed is 1.5 MPS (5 fps). Speed with an OBM is 4.5 MPS (15 fps).	The maximum current velocity with paddles is 1.5 MPS (5 fps). A 20 percent exit slope is desired. Three pumps and 11 paddles are included with each boat. OBMs must be requested separately.
Pneumatic, 3-man reconnaissance boat	L-series TOE provides— • 3 per combat engr co. • 18 per corps float bridge co. • 12 per div ribbon bridge co. • 18 per corps ribbon bridge co.	The boat is carried by backpack (1-man carry). The boat and backpack weigh 26 kg.	The boat can carry 3 soldiers with equipment or 306 kg of equipment.	Inflation time is 5 minutes with a pump. Paddle speed is 1.0 MPS (3 fps).	The maximum current velocity is 1.5 MPS (5 fps). One pump and 3 paddles are required per boat. The boat cannot be used without an OBM.
APC M113	J-series TOE provides— • 12 per engr co of engr bn. • 1 per inf co (mech) (BIFV). • 3 per inf co (mech) (M113). • 9 per armored engr co (ERI).	The APC— • Is self-propelled. • Is a Class 13 vehicle.	The APC can carry 12 soldiers with equipment.	Preparation time for swimming is 10 minutes. The APC is track-propelled in the water. Swimming speed is 1.6 MPS (5.3 fps). The APC can ford up to 1.5 meters (5 feet).	The maximum current velocity is 1.5 MPS (5 fps). $\text{Drift (meters)} = \frac{current}{1.6} x \left(\begin{array}{c} river\ width \\ [meters] \end{array} \right)$
BEB-SD	L-series TOE provides 14 per corps ribbon bridge co.	The boat is carried by one 5-ton bridge truck with a cradle or one medium-lift helicopter. The boat weighs 3,992 kg.	The boat can carry a 3-man crew and 12 soldiers with equipment or 1,996 kg of equipment.	Launch time from the cradle is 5 minutes.	The draft is— • 56 cm for normal operations. • 66 cm when fully loaded. • 122 cm for a launch from the cradle.

Table C-1. Equipment-characteristic chart (continued)

Equipment	Allocation	Transportation	Capabilities	Assembly/ Propulsion	Remarks/ Limitations
BIFV	J-series TOE provides— • 14 per inf co (mech) (BIFV). • 12 per cav troop of an ACR. • 19 per cav troop of a div cav squadron. • 80 per assault-boat team.	The BIFV— • Is self-propelled. • Is a Class 24 vehicle.	The BIFV can carry 10 solders with equipment.	Preparation time for swimming is 18 minutes. The BIFV is track-propelled in the water. Swimming speed is 2 MPS (6 fps). The BIFV can ford up to 1.1 meters (3.5 feet).	The maximum current velocity is 0.9 MPS (3 fps). $Drift \ (meters) = \dfrac{current}{2} x \left(\dfrac{river \ width}{[meters]} \right)$ $Drift \ (feet) = \dfrac{current}{6.6} x \left(\dfrac{river \ width}{[meters]} \right)$

Table C-2. Typical external loads

Equipment	Weight in Kilograms (pounds)	Remarks
M4T6 fixed spans 23 feet 4 inches, Class 100 30 feet 0 inches, Class 65 38 feet 4 inches, Class 35 45 feet 0 inches, Class 25	 5,851 (12,900) 7,076 (15,600) 8,528 (18,800) 9,480 (20,900)	Components are assembled in 8-foot 4-inch and 15-foot 0-inch increments. They may be transported in packages to reduce the load. Load class may be increased by varying the deck size.
Pneumatic assault boat	131 (290)	Boats are transported in a bundle or in an inflated mode.
27-foot BEB-SD	3,084 (6,800)/ 3,992 (8,800)	Boats are lifted in the bow-and-stem configuration.
M4T6 float-bridge components Float without deck Float with deck Two floats with partial deck	 3,039 (6,700) 5,307 (11,700) 7,666 (16,900	Loads are placed on the water or shore for further assembly.
Ribbon-bridge bays Interior bays End bays	 5,443 (12,000) 5,307 (11,700)	Bays are placed directly on water surfaces.

Figure C-1. Armored vehicle-launched bridge

Table C-3. Characteristic/operation of the AVLB

Allocation	Transportation	Emplacement	Capacity Class	Limitation/ Remarks
6 per engr co, ACR 6 per engr co, separate heavy brigade 4 per engr co, heavy division 4 per engr co, corps (mech)	The AVLB— • Is carried on a launcher (a modified M48A5 or an M60A1 chassis). • Weighs 15,000 kg (15 tons) (bridge only). The spare bridge is folded on a 25-ton lowbed trailer with a 10-ton tractor (usually consolidated at corps or theater level).	The AVLB— • Can be launched in 2 to 5 minutes by a buttoned-up 2-man crew. • Can be retrieved from either end. • Requires that one man be exposed to guide and connect while retrieving.	The total length of the AVLB is 19.2 meters (63 feet).* The AVLB is capable of holding a Class 60 vehicle across— • A 17.4-meter (57-foot) gap with unprepared abutments. • An 18.3-meter (60-foot) gap with prepared abutments.	The scissors launch requires 10 meters (32 feet) of overhead clearance. The maximum launch slope is— • 28 percent uphill. • 19 percent downhill. • 11 percent on a side slope.
NOTE: *For crossings on the AVLB that exceed MLC 60, see the current safety message.				

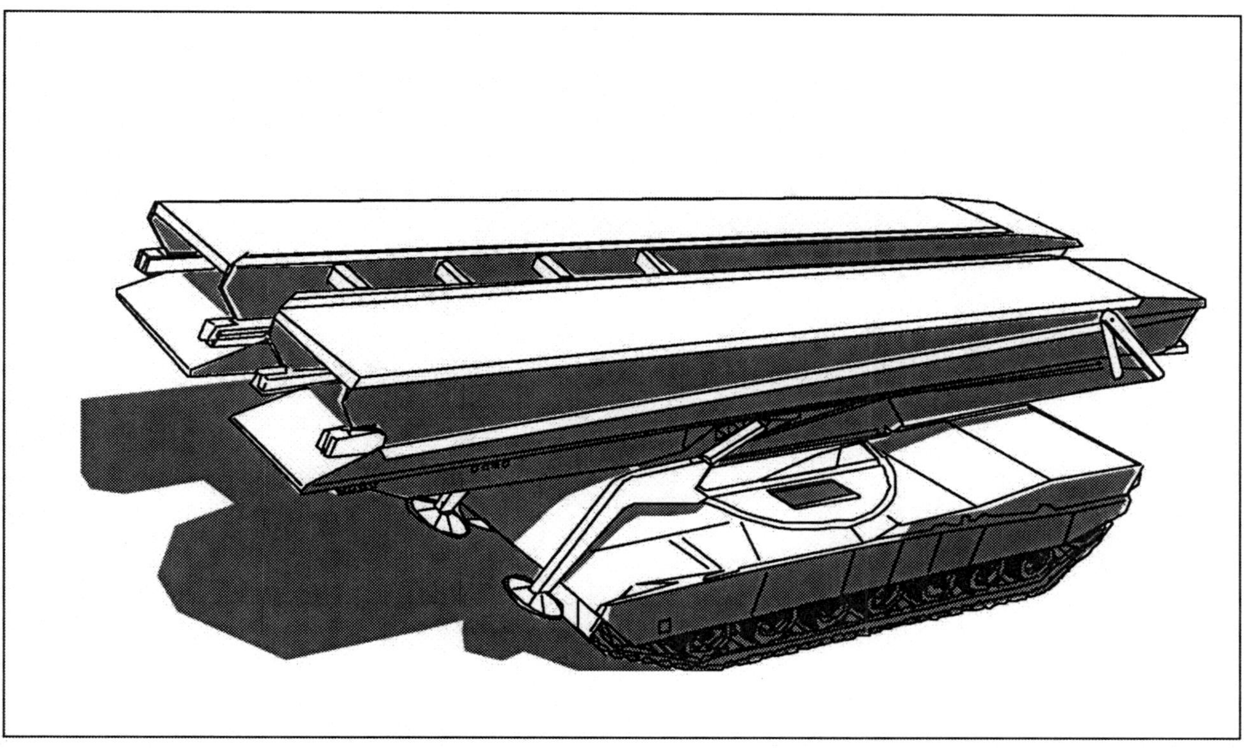

Figure C-2. Wolverine

Table C-4. Characteristics of the Wolverine

Allocation	Transportation	Emplacement	Capacity Class	Limitations/ Remarks
6 per engr co, ACR 6 per engr co, separate heavy bridge 4 per engr co, heavy division 4 per engr co, corps (mech)	The Wolverine— • Is carried on a launcher (a modified M1-series Abrams tank chassis.) • Weighs 12,500 kg (12.5 tons) (bridge only)	The Wolverine— • Can be launched in less than 5 minutes by a buttoned-up 2-man crew. • Can be retrieved from either end. • Can be recovered in less than 10 minutes.	The Wolverine— • Can hold a Class 70 vehicle. • Expands to a total length of 24 meters.	The launcher and bridge has a maximum speed of 83 kph. The Wolverine— • Can ford up to a depth of 122 cm (w/o a kit). • Will replace the AVLB one for one.

Figure C-3. Ribbon raft

Figure C-4. M4T6 raft

Table C-5. Launch restrictions

Characteristics	Free Launch	Controlled Launch	High-Bank Launch
Minimum depth of water required in centimeters (inches)	Ramp bay 112 (44) Interior bay 92 (36)[2]	76 (30)[1]	76 (30)[2]
Bank-height restrictions in meters (feet)	0-1.5 (0-5)	0	1.5 - 8.5 (5 - 28)
Bank-slope restrictions	0-30 percent	0-20 percent	Level the ground unless the front of the truck is restrained.

NOTES:

[1] This is the recommended water depth. The launch could technically be conducted in 43 centimeters (17 inches) of water.

[2] The launch is based on a 10 percent slope with the transporter backed into the water. The required water depth for a 30 percent slope with a 1.5-meter (5-foot) bank height is 183 centimeters (72 inches). Interpolate between these values when needed.

Table C-6. Allocation of ribbon bridge (L-series TOE)

Components	Per Corps Ribbon Company
Bridge platoons	2
Interior bays	30
Ramp bays	12
BEBs	15

NOTE: The longest bridge that can be constructed is 215 meters (705 feet).

Table C-7. Ribbon-raft design

Raft Types	Assembly Time in Minutes	Load Space in Meters (feet)	Classi-fication	Current Velocity in MPS (fps) and MLC							
				0-0.9 (0-3)	1.2 (4)	1.5 (5)	1.75 (6)	2 (7)	2.5 (8)	2.7 (9)	3 (10)
3 bays (2 ramps/ 1 interior)	8	6.7 (22)	L	45	45	45	40	40	35	30	25
			C	45	45	35	25	15	10	0	0
4 bays (2 ramps/ 2 interiors)	12	13 (44)	L	70	70	70	60	60	60	55	45
			C	60	60	60	55*	40*	30*	15*	0
5 bays (2 ramps/ 3 interiors)	15	20.1 (66)	L	75	75	75	70	70	70	60	60
			C	75	70	70	70*	60*	50*	25*	0
6 bays (2 ramps/ 4 interiors) wheeled/ tracked	20	26.8 (88)	L	96	96	96	96	96	96	70	70
				80	80	80	70	70	70	70	70
			C	96	96	96	70	70	55	30	
				70	75	70	70	70	55	30	0

NOTES:
1. When determining raft classification, L refers to the longitudinal rafting and C refers to conventional rafting.
2. If the current's velocity in the loading/unloading area is greater than 1.5 MPS (5 fps), then conventional rafting must be used.
3. The roadway width of a ribbon raft is 4.1 meters (13 feet 5 inches).
4. The draft of a fully loaded ribbon raft is 61 centimeters (24 inches).
5. Vehicles should only be loaded on the interior bays.
6. Each raft requires a minimum of two BEBs for propulsion.
7. The assembly time for a raft increases by 50 percent at night.
*Three BEBs are required for conventional rafting of 4, 5, or 6 bay rafts in current velocities greater than 1.5 MPS (5 fps).

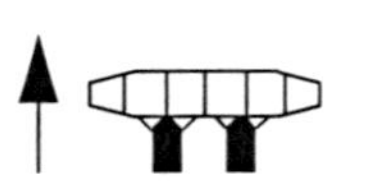
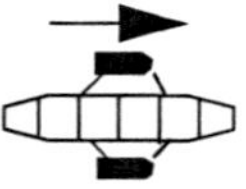

Conventional   Longitudinal

Table C-8. Raft-crossing capabilities

River Width		Minutes per Round Trip	Rounds Trips per Hour	Number of Rafts per Centerline
Feet	Meters			
246	75	7	8	1
328	100	8	7	1
410	125	9	6	1
492	150	10	6	2
610	188	11	5	2
738	225	12	5	2
861	263	14	4	3
964	300	16	3	3
1,148	350	18	3	4
1,312	400	20	3	5
1,476	450	22	2	5
1,640	500	24	2	5
1,968	600	26	2	6
2,296	700	29	2	6
2,824	800	32	1	6
2,952	900	35	1	6
3,280	1,000	38	1	6
3,808	1,100	41	1	6
3,936	1,200	45	1	6

NOTES:
1. **This table is valid for ribbon and M4T6 rafts in current velocities up to and including 1.5 MPS (5 fps). This data is based on the use of crews under ideal conditions.**
2. **Round-trip times include the times required to load and unload the raft.**
3. **Crossing times will take 50 percent longer at night.**
4. **If the river width falls between 2 columns, use the value found in the next higher column.**

Table C-9. M4T6-raft design and classification

Raft Types	Assembly Times	Crossing Types	Load Space in Meters (feet)	Current Velocity in MPS (fps) and MLC			
				1.5 (5)	2 (7)	2.5 (8)	3.5 (11)
4-float raft: 5 bridge trucks 2 BEB-SD 1 platoon	2 1/4 hours (when preassembled, 1 1/2 hours)	Normal	15.7 (51.6) (wheeled/ tracked)	50 55	45 50	40 45	30 35
		Reinforced	11.6 (38.3) (wheeled/ tracked)	50 55	50 55	45 50	35 40
5-float raft: 6 bridge trucks 2 BEB-SD 1 platoon	3 hours (when preassembled, 1 1/2 hours)	Normal	20.3 (66.6) (wheeled/ tracked)	55 60	50 55	45 50	35 40
		Reinforced	15.2 (50) (wheeled/ tracked)	60 65	60 65	55 60	45 50
6-float raft: 7 bridge trucks 2 BEB-SD 1 platoon	3 3/4 hours (when preassembled, 1 3/4 hours)	Reinforced	16.2 (53.3) (wheeled/ tracked)	65 70	65 70	65 70	45 50

NOTES:
1. For methods on constructing an M4T6 raft, refer to *TC 5-210*.
2. The roadway width for an M4T6 raft is 4.2 meters (13 feet 10 inches).
3. The draft of a fully loaded M4T6 raft is 66 centimeters (29 inches).
4. The assembly time for a raft increases by 50 percent at night.

Figure C-5. Ribbon bridge

The number of interior bays =

$$\frac{Gap\ (meters) - 14}{6.7}$$

or

$$\frac{Gap\ (feet) - 45}{22}$$

NOTES:
1. Two ramp bays are required for all ribbon bridges.
2. During daylight hours, a ribbon bridge can be constructed at the rate of 200 meters (600 feet) per hour and during nighttime hours, at the rate of 133 meters (437 feet) per hour.
3. Two hundred vehicles per hour, with 30-meter spacing at 16 kilometers per hour, can cross the bridge.

Figure C-6. Bridge design

Table C-10. Determination of bridge classification

Crossing Types	Current Velocity in MPS (fps) and MLC							
	0-0.9 (0-3)	1.2 (4)	1.5 (5)	1.75 (6)	2 (7)	2.5 (8)	2.7 (9)	3 (10)
Normal:								
wheeled	96	96	96	96	82	65	45	30
tracked	75	75	70	70	80	60	45	30
Caution:								
wheeled	105	105	100	100	96	75	50	35
tracked	85	85	80	80	80	65	50	35
Risk:								
wheeled	110	110	105	105	100	82	65	40
tracked	100	195	90	90	90	75	65	40

Table C-11. Number of boats needed for anchorage of a ribbon bridge

Current Velocity in MPS (fps)	Number of Boats: Number of Bridge Bays
0 to 2.0 (0 to 6.5)	1:6
2.0 to 2.6 (6.5 to 8.5)	1:3
2.7 (9)	1:2
Over 2.7 (over 9)	Bridge must be anchored using an overhead cable system.
NOTE: Anchorage of ribbon bridges is normally accomplished by tying BEBs to the downstream side of the bridge. The number of boats required is shown in the table.	

1. The corps float bridge company (M4T6) includes six sets of M4T6s and six BEBs.

2. One set provides—
- 43 meters (41 feet) of normal bridging
  or
- 29 meters (96 feet) of reinforced bridging
  or
- One 4-float normal raft
  or
- One 5-float normal raft
  or
- One 4-float reinforced raft and one 5-float reinforced raft
  or
- One 6-float reinforced raft.

3. The M4T6 is normally transported using a 5-ton bridge truck. One bay of a disassembled bridge can be loaded on one 5-ton truck. Bays can also be preassembled and flown to the river using medium-lift helicopters.

Figure C-7. Allocation and transportation factors
for the M4T6 bridge

Table C-12. Bridge classification for the M4T6

Crossing Types	Current Velocity in MPS (fps) and MLC			
	1.5 (5)	2 (7)	2.5 (8)	3.5 (11)
Normal				
Normal: wheeled	45	40	35	25
tracked	55	50	45	30
Caution: wheeled	58	54	49	35
tracked	59	55	51	37
Risk: wheeled	66	62	59	43
tracked	67	63	60	45
Reinforced				
Normal: wheeled	75	70	65	27
tracked		75	70	30
Caution: wheeled	80	79	73	43
tracked				45
Risk: wheeled	90	90	87	59
tracked				60

Table C-13. Determination of site and personnel requirements for the M4T6 bridge

Length for Normal Assembly in Meters (feet)	Units Needed for Assembly	Number of Assembly Sites	Time in Hours
45.5 (150)	1 company	2	4
61 (200)	1 company	2	5
76 (250)	1 company	2	6
91.5 (300)	2 companies	3	4
106.5 (350)	2 companies	3	5
122 (400)	2 companies	4	5.5
152 (500)	2 companies	5	6
183 (600)	3 companies	6	4
213 (700)	3 companies	6	5 to 7
244 (800)	3 companies	6	6 to 8
305 (1,000)	3 companies	6	7 to 10
366 (1,200)	3 companies	6	8 to 12

NOTES:
1. For methods on constructing an M4T6 bridge, refer to TC 5-210.
2. The construction time for a reinforced bridge should be increased by 50 percent.
3. The construction time for bridges should be increased by 50 percent at night.
4. The draft of an M4T6 bridge is 101.6 centimeters (40 inches).

Figure C-8. MGB

1. The corps MGB company includes—
   • Four bridge sets.
   • Two reinforcement sets.
   • Two erection sets.
2. Each MGB set requires seven 5-ton dump trucks and seven 4-ton bolster trailers for transportation.
3. Each link reinforcement set requires one 5-ton dump truck and one 4-ton bolster trailer for transportation.
4. Each erection set requires one 5-ton dump truck and one 4-ton bolster trailer for transportation.

Figure C-9. Allocation and transportation factors for the MGB

Table C-14. Work parties for MGBs

Bridging Activity	Work Party
4 and 5 bay SSB (7.9 to 9.8 meters or 26 to 32 feet)	1 NCO and 8 personnel
6 through 12 bay SSB (11.6 to 15.2 meters or 38 to 50 feet)	1 NCO and 16 personnel
All DSBs Anchorage party (if required)	1 NCO and 24 personnel 8 personnel
All DSBs with a link reinforcement party	1 NCO and 8 personnel

Table C-15. SSB length and classification for the MGB

Bridge Length		Number of Bays	MLC
Feet	Meters		
26	7.9	4	70
32	9.8	5	70
38	11.6	6	40
44	13.4	7	30
50	15.2	8	30
56	17.1	9	24
62	18.9	10	20
68	20.7	11	16
74	22.6	12	16

Table C-16. Building times (good conditions) for the MGB

Bridge Types	Bridge Sizes	Daytime Hours	Nighttime Hours
Single story	5 bays	0.50	0.75
	8 bays	0.75	1.00
	12 bays	1.00	1.50
Double story without LRS	4 bays	0.75	1.25
	8 bays	1.00	1.50
	12 bays	1.50	2.00
	18 bays	1.75	2.75
	22 bays	2.00	3.00
Double story with LRS	13 bays	2.00	3.00
	18 bays	2.75	4.00
	22 bays	3.00	4.50

NOTES:
1. A 25- by 20-meter assembly site is required.
2. Only MBG company personnel are required for assembly/disassembly.
3. The assembly time for bridges should be increased by 20 percent for untrained troops and 30 percent for inclement weather.

Table C-17. DSB length and classification for the MGB

| Bridge Length | | 2E + Number of Bays | MLC | |
Feet	Meters		Without LRS	With LRS
37	11.3	1	70	--
43	13.1	2	70	--
49	14.9	3	70	--
55	16.8	4	70	--
61	18.6	5	70	--
67	20.4	6	70	--
73	22.3	7	70	--
79	24.0	8	70	--
85	26.9	9	70	--
91	27.7	10	70	--
97	29.6	11	70	--
103	31.4	12	70	--
109	33.2	13	50	70
115	35.1	14	50	70
121	36.9	15	40	70
127	38.8	16	40	70
133	40.5	17	30	70
139	42.5	18	30	70
145	44.2	19	24	70
151	46.0	20	24	70
157	47.9	21	20	70
163	49.7	22	16	70

Table 18. Estimated time for assembly of the MGB

Span in Meters (feet)	Construction Type								
	SS	DS	TS	DD	TD	DT	TT	DT	TT
	Construction by Manpower Only							One Crane	
12 (40)	1 1/2								
18 (60)	1 3/4	2							
24 (80)	2	2 1/2	3						
30 (100)	2 1/4	3	3 1/2	4 1/4					
37 (120)		3 1/2	4	5	6 3/4				
43 (140)		3 3/4	4 1/2	5 3/4	7 1/2	11 1/4		10 1/2	
49 (160)			5	6 1/4	8 1/2	13 1/4	19	11 3/4	16 1/4
55 (180)				7	9 1/2	14 3/4	21 1/4	13 1/4	18 1/4
61 (200)						16 1/4	24	14 1/2	20 1/2

NOTES:
1. A 20- by 30-meter assembly site is required.
2. In addition to bridge company personnel, an assembly crew is required.
3. For more information, see FM 5-34 or FM 5-277.

Figure C-10. HDSB

Description

- The launcher and bridge components are on an ICBT, which is a HEMTT-based vehicle.
- The HDSB provides fast support bridging for LOC (from the brigade to the communications zone).

Capabilities

- One set can bridge a 40-meter gap.
- The set can be constructed in 90 minutes by 14 soldiers.
- One set can be constructed into two 20-meter bridges.
- Bridge components are transportable by a CH-47.
- The MLC for a 40-meter bridge is 96 wheeled/70 tracked.
- The bridge can span a 40-meter gap without intermediate support.

Requirements

- The minimum depth of the gap can be no less than 2.7 meters below the height of the bank.
- Vehicles of MLC 100 wheeled/80 tracked can cross under a caution crossing.
- The weight load should not exceed 10,659 kg on the M1077 palletized-load-system flat rack.
- The bridge sections will have a service-life monitoring system.

Figure C-11. HDSB characteristics

Figure C-12. M2 Bailey bridge

Table C-19. Classes of the M2 Bailey bridge

Type of Construction	Rating	Span in Feet																		
		30	40	50	60	70	80	90	100	110	120	130	140	150	160	170	180	190	200	210
SS	N	30	24		20	20	16	12	8											
		30		24																
	C	42	36	33	30	24	20	16	12											
		37	34	31	29															
	R	47	40	36	33	30	24	19	14											
		42	38	35	32	30														
DS	N			75	75	60	50	40	30	20	16	12	8							
				70	65	60	55	45	30											
	C			83	77	68	60	50	37	30	23	18	14							
				76	73	69	60	50	39	32										
	R			88	85	78	66	55	42	34	27	21	17							
				84	79	75	64	55	44	36	30									
TS	N						85	65	50	35	30	20	16	12	8	4				
							80	65	55	40	35									
	C						95	74	57	47	38	31	24	18	15	10				
							90	75	60	49	41	33								
	R						100*	82	64	52	43	35	29	22	17	13				
							90*	82	66	54	45	38	31							
DD	N								80	65	45	35	30	24	16	12	8			
									80	70	55	45	35							
	C								86	72	57	47	39	32	25	19	15			
									90	76	61	50	42	35						
	R								96	80	64	53	44	36	30	24	18			
									90	83	68	56	48	40	33					
TD	N									90	75	55	45	35	30	20	16	12		
										90*	80	60	55	45	35					
	C									100*	83	65	57	47	37	31	24	18		
										90*	90*	72	62	51	41	34				
	R									100*	91	74	64	54	45	37	29	22		
										90*	90*	80	70	58	48	40	32			
DT	N											70	70	60	55	45	35	30	20	16
												80	70	60	55	50	45	35		
	C											80	80	77	69	57	48	39	32	25
												90*	90*	85	78	64	58	43	36	
	R											90	88	85	80	64	55	46	38	31
												90*	90*	90*	89	74	60	51	43	35
TT	N														80	70	55	45	35	24
															75	70	60	55	40	
	C														100	80	66	59	48	38
															90*	90*	75	66	52	43
	R														100*	90	77	68	55	46
															90*	90*	87	77	62	51

Notes: N = Normal   C = Caution   R = Risk
1. Upper figure represents wheeled load class.
2. Lower figure represents tracked load class.
* Limited by roadway width.

$$\text{Class (wheeled)} = \frac{T^2 \times C \times S}{25}$$

$$\text{Class (tracked)} = \frac{T^2 \times C \times S}{20}$$

Legend:
T = Ice thickness in inches (see *Table C-20*)
C = Color factor (see *Table C-21, page C-26*)
S = Strength factor (see *Table C-22, page C-26*)

Figure C-13. Determining the class of ice

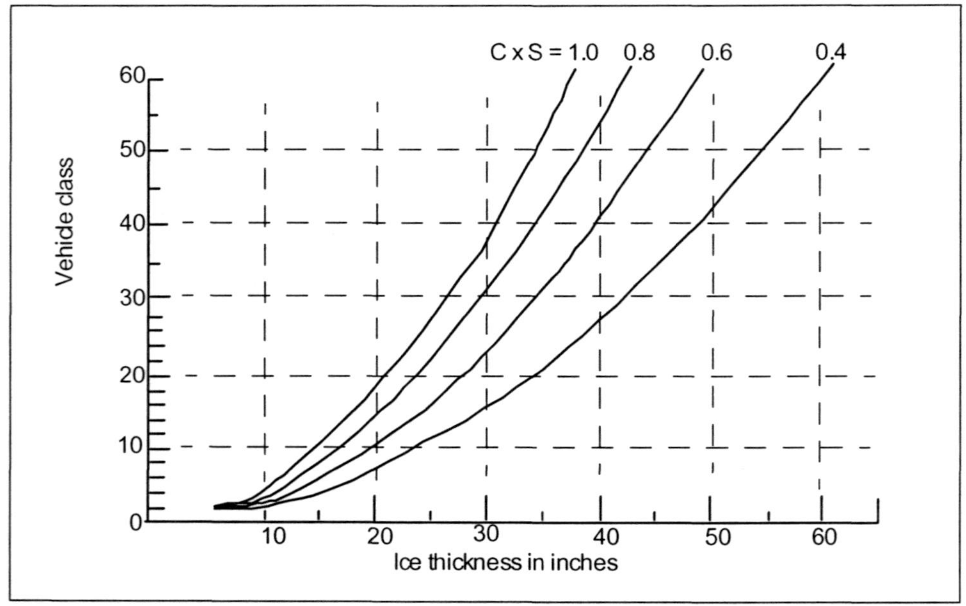

Figure C-14. Required ice thickness for wheeled vehicles

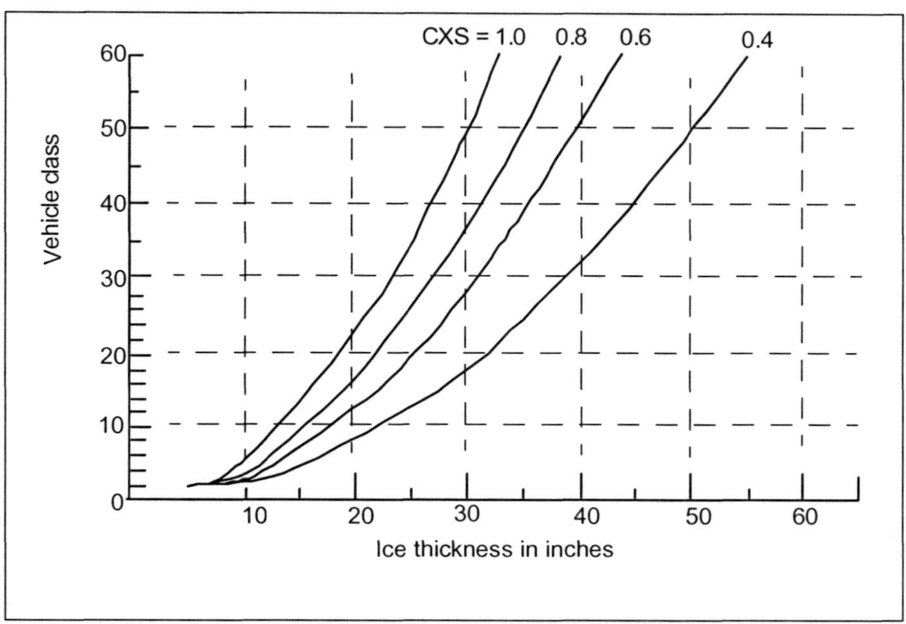

Figure C-15. Required ice thickness for tracked vehicles

Table C-20. Ice-depth requirements

Personnel	Ice-Thickness Requirements in Inches		
	Strong **C=1, S=1**	**Medium** **C=0.8, S=0.8**	**Weak** **C=0.7, S=0.6**
On skis	1.5	2	3
In a file formation with 2-meter intervals	3	4	5
On snowmobiles	3	4	5

Table C-21. Color factor

Factor	Characteristics
C = 1	Ice is clear (transparent)
C = 0.9	Ice is semiclear
C = 0.8	Ice is white
C = 0.7	Ice is discolored (stained brown or yellow)

Table C-22. Strength factor

Factor	Characteristics
S = 1	Ice is solid, and temperatures have remained at or below freezing for the previous week.
S = 0.9	Ice is solid, and temperatures have been above freezing during the day but drop below freezing during the night.
S = 0.8	Ice is solid, and water is running on the surface from runnoff or overflow.
S = 0.7	Ice is not solid, and water or air pockets are found in between layers of ice.
S = 0.6	An air pocket is under the ice, so the ice is not floating on the water underneath.

Table C-23. Method for determining vehicle distance

Vehicle Class (wheeled or tracked)	Required Ice Thickness in Centimeters	Distance Between Vehicles in Meters (about 100 x ice thickness [in cm])
1	11	11
2	15	15
3	18	18
4	21	21
5	23	23
10	33	33
15	40	40
20	46	46'
25	51	51
30	56	56
35	61	61
40	65	65
50	72	72
60	79	79
70	85	85
80	91	91

Before using the table, see remarks below:

1. If the air temperature has been above freezing for more than 6 of the past 24 hours, multiply the vehicle class by 1.3 to obtain the required ice thickness. If the air temperature stays above freezing for 2 hours or more, the ice starts to lose strength, and the table no longer represents safe conditions. A rapid and unusually large temperature drop causes the ice to become brittle, and travel may not be safe for a period of 24 hours.
2. For the distance required between two vehicles of different classes, use the distance required for the higher class.
3. If you plan to park for extended periods, multiply the vehicle class by 2 to obtain the required ice thickness and maintain at least the original distance requirements. Drill a hole through the ice near the vehicle, and move if the ice begins to flood.
4. The ice must have water support. Be very careful close to the shore. Very often the water level will drop after freeze-up. When this happens, the ice close to the shore may no longer have water support.
5. Cracks are either dry or wet. If dry, they do not penetrate ice cover and can be ignored. If wet, multiply the vehicle class by 2 to obtain the required ice thickness, and try to drive straight across the cracks (avoid going parallel to wet cracks).

Glossary

AA	assembly area
AAG	Army artillery group
ACE	M9 armored combat earthmover
ACR	armored cavalry regiment
AD	air defense
ADA	air-defense artillery
ADC	assistant division commander
AFB	assault float bridge
alt	alternate
APC	armored personnel carrier
armd	armored
AT	antitank
attn	attention
AVLB	armored vehicle-launched bridge
bde	brigade
BEB	bridge-erection boat
BIFV	Bradley infantry fighting vehicle
BMAIN	brigade main CP
bn	battalion
BTAC	brigade tactical CP
C^2	command and control
CAC	crossing-area commander

CAE	crossing-area engineer
CAS	close air support
cav	cavalry
cbt	combat
CCIR	commander's critical information requirements
CFC	crossing-force commander
CFE	crossing-force engineer
cm	centimeter(s)
co	company
COA	course of action
const	construction
CP	command post
CSC	crossing-site commander
CSS	combat service support
DA	Department of the Army
DAG	division artillery group
DD	double double
div	division
DMAIN	division main CP
DREAR	division rear CP
DS	double single
DSB	double-story bridge
DT	double triple
DTAC	division tactical CP

DTO	division transportation officer
EA	engagement area
EEP	engineer equipment park
engr	engineer
ERI	engineer restructure initiative
ERP	engineer regulating point
FA	field artillery
FEBA	forward edge of the battle area
FM	field manual
FO	forward observer
fps	foot (feet) per second
FRAGO	fragmentary order
FSB	forward support battalion
FSCL	fire-support coordination line
G2	Assistant Chief of Staff, G2 (Intelligence)
G3	Assistant Chief of Staff, G3 (Operations and Plans)
G4	Assistant Chief of Staff, G4 (Logistics)
GLLD	ground-laser location indicator
HDSB	heavy dry-support bridge
HEMTT	heavy-expanded mobility tactical truck
HIMAD	high-to-medium-altitude AD
HMMWV	high-mobility multiwheeled vehicle
HQ	headquarters
ICBT	improved common bridge transporter

IDP	initial-delay position
inf	infantry
int	intermediate
IPB	intelligence preparation of the battlefield
kg	kilogram(s)
km	kilometer(s)
kph	kilometer(s) per hour
LBE	load-bearing equipment
LOA	limit of advance
LOC	lines of communication
LRS	link reinforcement set
LTR	light tactical raft
LZ	landing zone
M	meter(s)
MCWP	Marine Corps Warfighting Publication
mech	mechanized
METT-T	mission, enemy, terrain, troops, and time available
MGB	medium-girder bridge
MLC	military load class
MO	Missouri
MOPP	mission-oriented protective posture
MP	military police
mph	mile(s) per hour
MPS	meter(s) per second

MRBC	multirole bridge company
MSI	multispectral imagery
MTOE	modified table of organization and equipment
NA	not applicable
NATO	North Atlantic Treaty Organization
NBC	nuclear, biological, chemical
NCO	noncommissioned officer
obj	objective
OBM	outboard motor
OCOKA	observation, cover and concealment, obstacles, key terrain, and avenues of approach
OCONUS	outside continental US
OPFOR	opposing force
OPORD	operation order
OPSEC	operations security
PFS	pipe fascines system
PIR	priority intelligence requirements
PL	phase line
PLS	palletized load system
prep	preparation
RAG	regimental artillery group
RB	ribbon bridge
RB15	rubber boat 15
RL	release line

RP	release point
rpm	revolution(s) per minute
RTO	radio telephone operator
S2	Intelligence Officer (US Army)
S3	Operations and Training Officer (US Army)
S4	Supply Officer (US Army)
SAW	squad automatic weapon
SD	shallow draft
sep	separate
SHORAD	short-range AD
SOP	standing operating procedure
SP	start point
SS	single single
SSB	single single bridge
STANAG	Standardization Agreement
TAC	tactical CP
TC	training circular
TCP	traffic-control post
TD	triple double
TF	task force
TM	technical manual
TOE	table(s) of organization and equipment
TOW	tube-launched, optically tracked, wire-guided
TRADOC	United States Army Training and Doctrine Command

TS	triple single
TT	triple triple
TTP	tactics, techniques, and procedures
US	United States
veh	vehicle
w/o	without
WO	warning order
XO	executive officer

References

SOURCES USED

These are the sources quoted or paraphrased in this publication.

Army Publications

FM 5-34. *Engineer Field Data.* 14 September 1987.

FM 5-250. *Explosives and Demolitions.* 15 June 1992.

FM 5-277. *M2 Bailey Bridge.* 9 May 1986.

FM 34-130. *Intelligence Preparation of the Battlefield.* 8 July 1994.

FM 71-3. *The Armored and Mechanized Infantry Brigade.* 8 January 1996.

FM 71-100. *Division Operations.* 28 August 1996.

FM 90-4. *Air Assault Operators.* 16 March 1987.

FM 90-13-1. *Combined Arms Breaching Operations.* 28 February 1991.

FM 100-5. *Operations.* 14 June 1993.

FM 100-15. *Corps Operations.* 29 October 1996.

FM 101-5. *Staff Organization and Operations.* 25 May 1984.

FM 101-5-1. *Operational Terms and Symbols.* 21 October 1985.

TC 5-210. *Military Float Bridging Equipment.* 27 December 1988.

TRADOC Pamphlet 350-14. *Heavy Opposing Force (OPFOR) Tactical Handbook (S&I by USACAC).* 15 September 1994.

TRADOC Pamphlet 350-16. *Light Opposing Force (OPFOR) Tactics Handbook (S&I by USACAC).* 15 September 1994.

Standardization Agreements

STANAG 2395. *Opposed Water Crossing Procedures (Edition 1).* 27 October 1994.

DOCUMENTS NEEDED

This document must be available to the intended users of this publication.

DA Forms

DA Form 2028. *Recommended Changes to Publications and Blank Forms.* 1 February 1974.

Index

mission, enemy, terrain, troops, and time available (METT-T), 5-2, 5-5, 5-7, 5-8, 8-3
mission-oriented protective posture (MOPP), 2-7
mock-up raft, 9-2
MOPP. *See* mission-oriented protective posture (MOPP).
Morse Code, 8-8
MRBC. *See* multirole bridge company (MRBC).
MSI. *See* multispectral imagery (MSI).
multirole bridge company (MRBC), 9-8, 9-9
multirole-bridge-company (MRBC) concept, 9-10
multispectral imagery (MSI), 2-5

O

observation, cover and concealment, obstacles, key terrain, and avenues of approach (OCOKA), 2-2
obstructions, 2-3
 man-made underwater, 2-3
 mud banks, 2-3
 natural underwater, 2-3
 rocks, 2-3
 sand, 2-3
 vegetation, 2-3
OCOKA. *See* observation, cover and concealment, obstacles, key terrain, and avenues of approach (OCOKA).
Operations and Training Officer (US Army) (S3), 3-14, 4-4
operations security (OPSEC), 1-4, 6-7
OPSEC. *See* operations security (OPSEC).

P

PFS. *See* pipe fascines system (PFS).
pipe fascines system (PFS), 1-2
PIR. *See* priority intelligence requirements (PIR).
pneumatic, 15-man assault boat, C-4
preparation team, 8-9
priority intelligence requirements (PIR), 2-6

R

raft
 maintenance, 7-13

operations, 7-12, 7-13, 9-2
 refueling, 7-13
 site, 7-9, 7-11, 7-12
rafts, 7-9
 M4T6, C-2, C-8, C-12
 ribbon, C-2, C-8, C-10
RAGs. *See* regimental artillery groups (RAGs).
reconnaissance
 far-shore, 8-7, 8-8
 ground, 7-5
 nearshore, 8-10
 team, 8-7, 8-8
regimental artillery groups (RAGs), 2-7
rise, 7-6
river-crossing categories
 caution, C-1
 normal, C-1
 risk, C-1
river-crossing organization, 1-3
 assault force, 1-3, 1-5, 3-1, 7-9, 8-4, 8-9
 bridgehead force, 1-3, 1-5, 3-1
 maneuver-support force, 1-3, 1-5, 3-1, 4-1
RL. *See* control measures, release line (RL).
rubber boats, 8-2
run, 7-6

S

S2. *See* Intelligence Officer (US Army) (S2).
S3. *See* Operations and Training Officer (US Army) (S3).
S4. *See* Supply Officer (US Army) (S4).
safety boat, 7-11
SHORAD. *See* short-range air-defense (SHORAD).
short-range air-defense (SHORAD), 5-4, 5-6, 5-9
slopes and degrees, 7-5, 7-6
smoke production, 8-16
smoke, 8-4
staging area, 3-9, 3-10, 3-12, 9-2
Supply Officer (US Army) (S4), 3-3
support force, 8-2
swimming operations, 9-4

T

TAC. *See* tactical command post (TAC).

Printed in the United States
113015LV00004B/119/A